计算机技术
开发与应用丛书

Unity编辑器开发与拓展

张寿昆 ◎ 著

清华大学出版社

北京

内 容 简 介

本书系统地讲解 Unity 编辑器开发工作中常用的类与方法,以基础知识为核心,结合实战案例,引导读者渐进式地学习与 Unity 编辑器开发相关的技术知识。

本书共 7 章,依次介绍编辑器开发的基础、如何自定义编辑器功能菜单、如何自定义检视面板、如何自定义编辑器窗口、如何定制编辑器的外观样式、如何使用编辑器辅助开发调试,以及在编辑器环境中进行数据与资产管理等相关内容。

本书既适合初学者入门,对有多年工作经验的开发者也具有参考价值。

本书封面贴有清华大学出版社防伪标签,无标签者不得销售。
版权所有,侵权必究。举报: 010-62782989,beiqinquan@tup.tsinghua.edu.cn。

图书在版编目(CIP)数据

Unity 编辑器开发与拓展 / 张寿昆著. —北京: 清华大学出版社,2024.4
(计算机技术开发与应用丛书)
ISBN 978-7-302-66074-3

Ⅰ.①U… Ⅱ.①张… Ⅲ.①程序设计 Ⅳ.①TP311.1

中国国家版本馆 CIP 数据核字(2024)第 072131 号

责任编辑:赵佳霓
封面设计:吴　刚
责任校对:郝美丽
责任印制:宋　林

出版发行:清华大学出版社
网　　址:https://www.tup.com.cn,https://www.wqxuetang.com
地　　址:北京清华大学学研大厦 A 座　　邮　编:100084
社 总 机:010-83470000　　邮　购:010-62786544
投稿与读者服务:010-62776969,c-service@tup.tsinghua.edu.cn
质 量 反 馈:010-62772015,zhiliang@tup.tsinghua.edu.cn
课 件 下 载:https://www.tup.com.cn,010-83470236
印 装 者:三河市龙大印装有限公司
经　　销:全国新华书店
开　　本:186mm×240mm　　印　张:17.75　　字　数:402 千字
版　　次:2024 年 5 月第 1 版　　印　次:2024 年 5 月第 1 次印刷
印　　数:1～2000
定　　价:69.00 元

产品编号:105505-01

前 言
PREFACE

Unity 作为一款强大的游戏开发引擎，其编辑器功能的可扩展性和可定制性一直是开发者所青睐的。作者在最初学习时，苦于没有系统的学习资料，只能在不断摸索中积累经验。本次写作的目的是希望本书能够为想要学习编辑器开发方向内容的开发者提供详细的学习资料。

本书第 1 章介绍了编辑器开发的基础，包括绘制各种类型的编辑器元素，以及如何进行编辑器布局。第 2~4 章分别介绍了如何自定义编辑器功能菜单、检视面板和编辑器窗口。第 5 章介绍了与编辑器外观相关的内容，包括皮肤、样式、图标和动画。第 6 章介绍了 Gizmos、Handles 两个可视化辅助工具。第 7 章介绍了编辑器环境中数据与资产管理的相关内容。

通过本书的学习，读者将能够轻松地定制工作所需的编辑器工具，提高工作效率，优化工作流程。本书在写作过程中使用的 Unity 版本为 2020.3.16f1c1，因为不同版本的 API 可能会略有不同，因此建议读者在学习过程中使用相同的版本。扫描目录上方的二维码可下载本书源代码。

在写作过程中，作者得到了家人和朋友的帮助，在此表示感谢。同时，感谢清华大学出版社赵佳霓编辑的细心指导。

限于作者知识水平，书中难免存在不妥之处，欢迎读者批评指正。

张寿昆

2024 年 3 月

目录
CONTENTS

本书源代码

第 1 章 编辑器开发基础 ··············· 1
  1.1 绘制编辑器元素 ··················· 2
    1.1.1 文本 ························ 2
    1.1.2 按钮 ························ 4
    1.1.3 开关 ························ 6
    1.1.4 输入框 ······················ 6
    1.1.5 下拉列表 ···················· 8
    1.1.6 滑动条 ······················ 9
    1.1.7 折叠栏 ····················· 10
  1.2 编辑器布局 ······················ 11
    1.2.1 水平与垂直布局 ············· 11
    1.2.2 GUI 中的滚动列表 ··········· 13
    1.2.3 GUI 元素和布局的大小 ······· 14
    1.2.4 GUI 元素之间的间隙 ········· 18

第 2 章 自定义功能菜单 ·············· 20
  2.1 MenuItemAttribute ··············· 20
    2.1.1 自定义 Unity 顶部的功能菜单 ···· 20
    2.1.2 自定义 Hierarchy 窗口右键功能菜单 ···· 28
    2.1.3 自定义 Project 窗口右键功能菜单 ···· 33
    2.1.4 自定义组件下拉列表功能菜单 ···· 37
  2.2 ContextMenuAttribute ············ 40
  2.3 ContextMenuItemAttribute ········ 41

第 3 章 自定义检视面板 ·············· 43
  3.1 创建自定义编辑器类 ·············· 43
    3.1.1 如何自定义检视面板中的 GUI 内容 ···· 44
    3.1.2 如何检测和应用修改 ········· 46

- 3.1.3 编辑器操作的撤销与恢复 ... 48
- 3.1.4 实现 DoTween 动画参数的编辑 ... 53
- 3.1.5 如何自定义预览窗口 ... 58
- 3.1.6 扩展默认组件的检视面板 ... 67
- 3.2 PropertyDrawer ... 74
  - 3.2.1 内置的 PropertyDrawer ... 74
  - 3.2.2 内置的 DecoratorDrawer ... 78
  - 3.2.3 如何创建自定义 PropertyDrawer ... 79

第 4 章 自定义编辑器窗口 ... 86

- 4.1 如何创建新的编辑器窗口 ... 86
  - 4.1.1 打开新创建的编辑器窗口 ... 86
  - 4.1.2 定义编辑器窗口中的 GUI 内容 ... 87
  - 4.1.3 如何创建弹出窗口 ... 89
  - 4.1.4 开发备忘录 ... 93
  - 4.1.5 Protobuf 通信协议文件编辑器 ... 105
  - 4.1.6 ScriptableWizard ... 117
- 4.2 如何扩展默认的编辑器窗口 ... 120
  - 4.2.1 扩展 Hierarchy 窗口 ... 120
  - 4.2.2 扩展 Project 窗口 ... 121
- 4.3 Game 窗口中的 GUI ... 123
  - 4.3.1 运行时控制台窗口 ... 125
  - 4.3.2 运行时层级窗口 ... 133
  - 4.3.3 运行时检视窗口 ... 137

第 5 章 编辑器外观 ... 151

- 5.1 GUI 皮肤 ... 151
- 5.2 GUI 样式 ... 154
- 5.3 GUI 图标 ... 159
- 5.4 GUI 动画 ... 163

第 6 章 可视化辅助工具 ... 165

- 6.1 Gizmos ... 165
  - 6.1.1 概述 ... 165
  - 6.1.2 常用函数 ... 169
  - 6.1.3 使用 Gizmos 辅助调试相机的避障功能 ... 176

6.2 Handles ···················· 181
6.2.1 概述 ···················· 181
6.2.2 常用函数 ···················· 182
6.2.3 实现一个路径编辑工具 ···················· 194

第 7 章 编辑器环境下的数据与资产管理 ···················· 207
7.1 EditorPrefs ···················· 207
7.2 AssetDatabase ···················· 209
7.2.1 资产管理 ···················· 209
7.2.2 AssetBundle 管理 ···················· 232
7.2.3 CustomPackage 管理 ···················· 244
7.3 ScriptableObject ···················· 245
7.4 AssetModificationProcessor ···················· 248
7.5 AssetPostprocessor ···················· 251
7.6 BuildPipeline ···················· 257
7.6.1 AssetBundle 构建工具 ···················· 258
7.6.2 应用程序批量构建工具 ···················· 266

第 1 章 编辑器开发基础

本章主要介绍在编辑器开发中常用的、基础的 API，例如如何绘制文本、按钮、开关、输入框、下拉列表等 GUI 元素，以及如何对这些 GUI 元素进行布局。

在此之前，需要先了解什么是编辑器脚本？它们是任何一段使用 UnityEditor 命名空间下类和方法的代码脚本，这些代码脚本存放于 Editor 特殊文件夹下，只能在编辑器环境中使用，并且在打包时不会被打进包体中。

Editor 文件夹可以在 Assets 根目录下，也可以在任何子目录下，例如 Assets/Editor、Assets/Scripts/Editor，在一个项目工程中可以有任意多个 Editor 文件夹。假如在 Editor 文件夹之外的脚本中使用 UnityEditor 中的类和方法，在工程执行 Build 命令时会遇到编译报错，导致打包失败。

例如，在 Assets/Scripts 目录中创建一个名为 Example.cs 的示例脚本组件，在该组件中调用了 EditorUtility 类中的方法，这个类是 UnityEditor 中的类，代码如下：

```
using UnityEngine;

public class Example : MonoBehaviour
{
    private void Start()
    {
        UnityEditor.EditorUtility.DisplayDialog("title", "message", "OK");
    }
}
```

然后在 Build Settings 窗口中单击 Build 按钮确认打包后，便会遇到编译报错的情况，如图 1-1 所示。

当然这个问题可以通过条件编译来规避，将使用了编辑器相关类和方法的代码放在被 UNITY_EDITOR 的宏所包含的代码中，使其仅在编辑器环境中生效，代码如下：

```
using UnityEngine;

public class Example : MonoBehaviour
{
    private void Start()
```

```
        {
#if UNITY_EDITOR
            UnityEditor.EditorUtility.DisplayDialog("title", "message", "OK");
#endif
        }
}
```

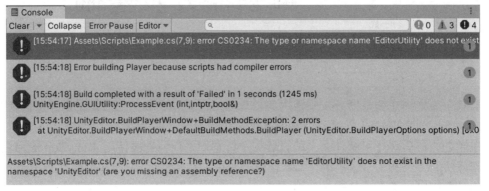

图 1-1　编译报错

1.1　绘制编辑器元素

编辑器是由若干 GUI 元素组成的，本节介绍如何绘制不同类型的 GUI 元素，为自定义组件 Example 创建一个编辑器类 ExampleEditor，以其为示例，介绍如何在组件的检视面板（Inspector）中绘制各种类型的 GUI 元素。

为自定义组件创建的编辑器称为自定义编辑器，自定义编辑器需要继承 Editor 类，并且需要为该类使用 CustomEditor 特性。重写 Editor 类中的虚方法 OnInspectorGUI()即可自定义组件的检视面板如何绘制，自定义编辑器的相关内容会在后续章节中详细介绍，本节主要介绍 GUI 元素的绘制。

1.1.1　文本

在自定义编辑器中可以通过 GUILayout 类中的 Label()方法或 EditorGUILayout 类中的 LabelField()方法来绘制文本，并且可以通过 GUIStyle 类型的参数指定文本字体的风格，例如小号字体、大号字体、加粗字体等。可以使用 Editor Styles 类中已有的一些字体风格，见表 1-1，也可以进行自定义，GUIStyle 的内容将在后续章节中进行详细介绍。

wrapped 类型字体样式的作用是当文本内容的长度大于 Label 标签的宽度时会进行换行显示，类似于 UGUI 中 Text 组件的 Horizontal Overflow 属性为 Wrap 类型。

在示例脚本中分别使用 EditorStyles 类中的这些字体样式绘制文本，看一下效果，代码如下：

表 1-1　EditorStyles 类中的字体样式

字　体　样　式	描　　述	字　体　样　式	描　　述
lable	默认的字体	wordWrappedLabel	包裹的字体
miniLabel	小号字体	linkLabel	链接样式的字体
largeLabel	大号字体	whiteLabel	白色字体
boldLabel	加粗字体	whiteMiniLabel	白色、小号字体
miniBoldLabel	小号、加粗字体	whiteLargeLabel	白色、大号字体
centeredGreyMiniLabel	居中、灰色、小号字体	whiteBoldLabel	白色、加粗字体
wordWrappedMiniLabel	包裹的、小号字体		

```csharp
//第1章/ExampleEditor.cs

using UnityEngine;
using UnityEditor;

[CustomEditor(typeof(Example))]
public class ExampleEditor : Editor
{
    public override void OnInspectorGUI()
    {
        base.OnInspectorGUI();
        Label();
    }
    private void Label()
    {
        GUILayout.Label("Hello world.", EditorStyles.label);
        GUILayout.Label("Hello world.", EditorStyles.miniLabel);
        GUILayout.Label("Hello world.", EditorStyles.largeLabel);
        GUILayout.Label("Hello world.", EditorStyles.boldLabel);
        GUILayout.Label("Hello world.", EditorStyles.miniBoldLabel);
        GUILayout.Label("Hello world.",
            EditorStyles.centeredGreyMiniLabel);
        GUILayout.Label("Hello world.",
            EditorStyles.wordWrappedMiniLabel, GUILayout.Width(50f));
        GUILayout.Label("Hello world.", EditorStyles.wordWrappedLabel);
        GUILayout.Label("Hello world.", EditorStyles.linkLabel);
        GUILayout.Label("Hello world.", EditorStyles.whiteLabel);
        GUILayout.Label("Hello world.", EditorStyles.whiteMiniLabel);
        GUILayout.Label("Hello world.", EditorStyles.whiteLargeLabel);
        GUILayout.Label("Hello world.", EditorStyles.whiteBoldLabel);
        //自定义字体样式：右对齐、字号为20
        GUILayout.Label("Hello world.", new GUIStyle()
        {
            alignment = TextAnchor.LowerRight,
```

```
            fontSize = 20
        });
    }
}
```

结果如图 1-2 所示。

图 1-2 文本

1.1.2 按钮

按钮是常见的 GUI 元素，因为大部分交互行为需要按钮来提供事件响应，按钮可以通过 GUILayout 类中的 Button() 方法进行绘制。

在使用 UGUI 中的 Button 组件时，可以通过调用其 onClick.AddListener() 方法来添加按钮的响应事件，那么在编辑器中如何为按钮添加单击时的回调事件呢？Button() 方法的返回值为布尔类型，它表示的是该按钮当前是否被按下，因此，将按钮的响应事件写在 if 代码块中即可，代码如下：

```
if (GUILayout.Button("Button"))
{
    Debug.Log("按钮被单击");
}
```

按钮同样可以通过 GUIStyle 类型参数设置其样式，EditorStyles 类中已有的按钮样式见表 1-2。

表 1-2　EditorStyles 类中的按钮样式

按 钮 样 式	描　　述	按 钮 样 式	描　　述
miniButton	相对较小的按钮	miniButtonLeft	一组按钮中最左侧的按钮
radioButton	开关样式的按钮	miniButtonMid	一组按钮中中间的按钮
toolbarButton	工具栏样式的按钮	miniButtonRight	一组按钮中最右侧的按钮

miniButtonLeft、miniButtonMid、miniButtonRight 样式可以组合使用，通常被用于水平方向排列的一组按钮中，左右两侧的按钮分别使用 miniButtonLeft、miniButtonRight 样式，miniButtonMid 样式在这组按钮数量大于 2 时被用于中间。

在示例脚本中分别使用 Editor Styles 类中的这些按钮样式绘制按钮，看一下效果，代码如下：

```
//第1章/ExampleEditor.cs
public override void OnInspectorGUI()
{
    base.OnInspectorGUI();
    Button();
}
private void Button()
{
    GUILayout.Button("Button1");
    GUILayout.Button("Button2", EditorStyles.miniButton);
    GUILayout.Button("Button3", EditorStyles.radioButton);
    GUILayout.Button("Button4", EditorStyles.toolbarButton);
    //水平方向布局
    GUILayout.BeginHorizontal();
    GUILayout.Button("Button5", EditorStyles.miniButtonLeft);
    GUILayout.Button("Button6", EditorStyles.miniButtonMid);
    GUILayout.Button("Button7", EditorStyles.miniButtonMid);
    GUILayout.Button("Button8", EditorStyles.miniButtonRight);
    GUILayout.EndHorizontal();
}
```

结果如图 1-3 所示。

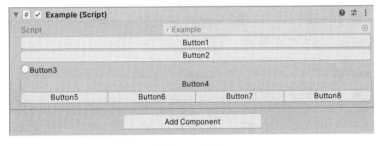

图 1-3　按钮

1.1.3 开关

开关控件用于布尔类型变量的交互修改，在自定义编辑器中可以通过 GUILayout 类或 EditorGUILayout 类中的 Toggle()方法来创建一个开关控件，方法的返回值也是布尔类型，表示开关的状态，因此，如果想要通过开关控件来交互修改一个布尔类型的变量，则可将该变量作为参数传入方法并使用该变量接收方法返回值，示例代码如下：

```
//第1章/ExampleEditor.cs

private bool boolValue1;
private bool boolValue2;

public override void OnInspectorGUI()
{
    base.OnInspectorGUI();
    Toggle();
}
private void Toggle()
{
    boolValue1 = GUILayout.Toggle(boolValue1, "Toggle1");
    boolValue2 = EditorGUILayout.Toggle("Toggle2", boolValue2);
}
```

结果如图 1-4 所示。

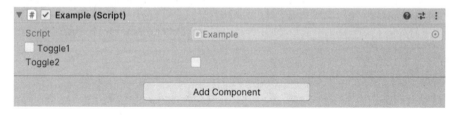

图 1-4　开关

很多交互控件的绘制和开关一样，既可以使用 GUILayout 类中的方法，也可以使用 EditorGUILayout 类中的方法，那么这两个类有什么区别？除了包含的方法略有差别外，它们的适用范围不同，GUILayout 类位于 UnityEngine 命名空间中，而 EditorGUILayout 类位于 UnityEditor 命名空间中，因此后者只能在 Editor 文件夹下的脚本中使用，而前者还经常被用于 MonoBehaviour 的 OnGUI()函数中，使用范围更广。

1.1.4 输入框

输入框的类型有多种，可能想要通过输入框来交互修改字符串类型的变量、整数类型的变量，甚至是 Unity 中向量类型的变量。不同类型的输入框及绘制方法见表 1-3。

表 1-3 输入框类型及绘制方法

类 型	作 用	方 法
TextField	用于字符串类型变量的交互修改	GUILayout.TextField()或 EditorGUILayout.TextField()
FloatField	用于浮点数类型变量的交互修改	EditorGUILayout.FloatField()
IntField	用于整型变量的交互修改	EditorGUILayout.IntField()
PasswordField	用于字符串类型变量的交互修改，与 TextField 不同的是，它不以明文进行显示，适用于密码	GUILayout.PasswordField()或 EditorGUILayout.PasswordField()
LongField	用于长整型变量的交互修改	EditorGUILayout.LongField()
Vector2Field	用于二维向量类型变量的交互修改	EditorGUILayout.Vector2Field()
Vector3Field	用于三维向量类型变量的交互修改	EditorGUILayout.Vector3Field()
Vector4Field	用于四维向量类型变量的交互修改	EditorGUILayout.Vector4Field()

在示例脚本中声明不同类型的变量并调用相应的方法绘制输入框，代码如下：

```
//第1章/ExampleEditor.cs
private string stringValue = "Hello World.";
private float floatValue = 10f;
private int intValue = 5;
private long longValue = 1;
private string passwordValue = "123456";
private Vector2 vector2Value = Vector2.zero;
private Vector3 vector3Value = Vector3.zero;
private Vector4 vector4Value = Vector4.zero;

public override void OnInspectorGUI()
{
    base.OnInspectorGUI();
    InputField();
}
private void InputField()
{
    stringValue = EditorGUILayout.TextField("StringValue", stringValue);
    floatValue = EditorGUILayout.FloatField("FloatValue", floatValue);
    intValue = EditorGUILayout.IntField("IntValue", intValue);
    longValue = EditorGUILayout.LongField("LongValue", longValue);
    passwordValue = EditorGUILayout.PasswordField(
        "PasswordValue", passwordValue);
    vector2Value = EditorGUILayout.Vector2Field(
        "Vector2Value", vector2Value);
    vector3Value = EditorGUILayout.Vector3Field(
        "Vector3Value", vector3Value);
    vector4Value = EditorGUILayout.Vector4Field(
```

```
    "Vector4Value", vector4Value);
}
```

结果如图1-5所示。

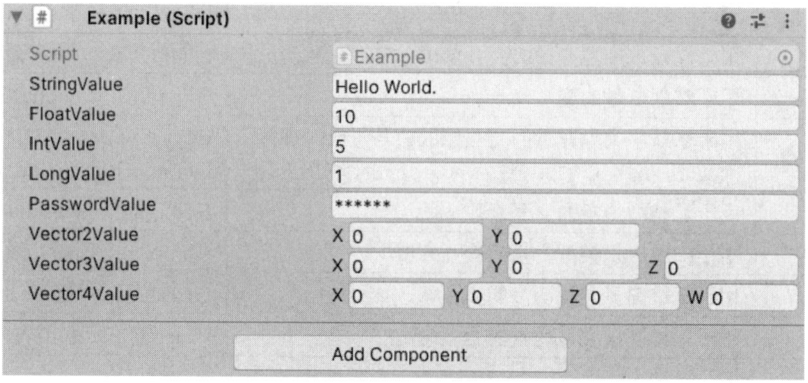

图1-5 输入框

1.1.5 下拉列表

下拉列表控件主要用于枚举类型变量的交互修改，控件的绘制通过调用 EditorGUILayout 类中的 EnumPopup()方法，需要注意的是，方法的返回值是 Enum 类型，因此会有类型转换的过程。除此之外，游戏物体的 Tag 标签、Layer 层级也可以通过下拉列表的形式进行选择，调用的方法分别是 EditorGUILayout 类中的 TagField()和 LayerField()，示例代码如下：

```
//第1章/ExampleEditor.cs

public enum ExampleEnum
{
   Enum1,
   Enum2,
   Enum3
}
private ExampleEnum enumValue = ExampleEnum.Enum1;

public override void OnInspectorGUI()
{
   base.OnInspectorGUI();
   Dropdown();
}
private void Dropdown()
{
   enumValue = (ExampleEnum)EditorGUILayout.EnumPopup(
      "EnumValue", enumValue);
   Selection.activeGameObject.tag = EditorGUILayout.TagField(
      "TagValue", Selection.activeGameObject.tag);
```

```
        Selection.activeGameObject.layer = EditorGUILayout.LayerField(
            "LayerValue", Selection.activeGameObject.layer);
}
```

注意：Selection.activeGameObject 表示用户在 Hierarchy 或 Project 窗口中所选中的游戏物体。

结果如图 1-6 所示。

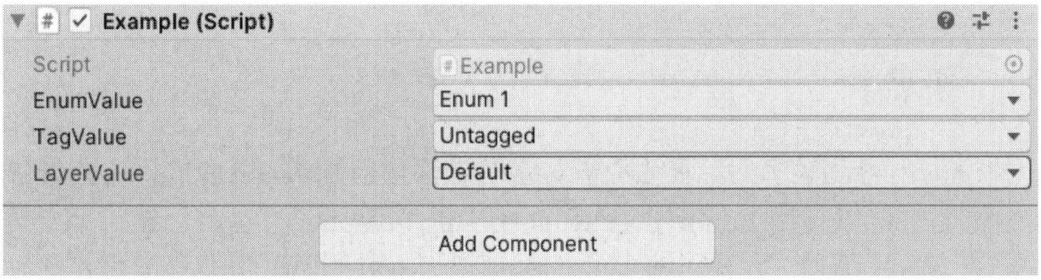

图 1-6 下拉列表

1.1.6 滑动条

滑动条适用于整数类型与浮点数类型变量的交互修改，与输入框不同的是，它可以限制变量的取值范围。整数类型的滑动条使用 EditorGUILayout 类中的 IntSlider() 方法绘制，浮点数类型的滑动条使用 EditorGUILayout 类中的 Slider() 方法绘制，变量的取值范围通过方法的参数进行设定。

在示例脚本中分别绘制一个整数类型的滑动条和一个浮点数类型的滑动条，取值范围分别设为[0, 5]和[0, 10]，代码如下：

```
//第1章/ExampleEditor.cs

private int intValue = 1;
private float floatValue = 3f;

public override void OnInspectorGUI()
{
    base.OnInspectorGUI();
    Slider();
}
private void Slider()
{
    intValue = EditorGUILayout.IntSlider("IntValue",
        intValue, 0, 5);  //取值范围为0~5
    floatValue = EditorGUILayout.Slider("FloatValue",
        floatValue, 0f, 10f);  //取值范围为0~10
}
```

结果如图 1-7 所示。

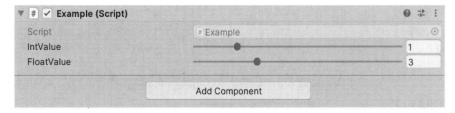

图 1-7 滑动条

1.1.7 折叠栏

当 GUI 中的内容比较多时，通常希望对它们进行分门别类，以便放到不同的折叠栏中。在自定义编辑器中可以使用 EditorGUILayout 类中的 Foldout()方法来创建一个折叠栏，方法的第 1 个参数为布尔类型，表示折叠栏的状态是否为打开，如果折叠栏的状态通过交互发生了变更，则可以通过方法的返回值获取。方法的第 2 个参数用于给折叠栏命名，第 3 个参数用于设置是否允许单击折叠栏的文本时切换折叠状态，示例代码如下：

```
//第1章/ExampleEditor.cs

private bool foldout1;
private bool foldout2;

public override void OnInspectorGUI()
{
   base.OnInspectorGUI();
   Foldout();
}
private void Foldout()
{
   foldout1 = EditorGUILayout.Foldout(foldout1, "折叠栏1", true);
   if (foldout1)
   {
      GUILayout.Label("Hello world.", EditorStyles.miniLabel);
      GUILayout.Label("Hello world.", EditorStyles.boldLabel);
      GUILayout.Label("Hello world.", EditorStyles.largeLabel);
   }
   foldout2 = EditorGUILayout.Foldout(foldout2, "折叠栏2", true);
   if (foldout2)
   {
      GUILayout.Button("Button1");
      GUILayout.Button("Button2");
      GUILayout.Button("Button3");
   }
}
```

结果如图 1-8 所示。

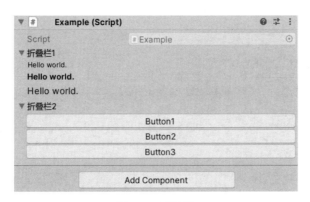

图 1-8 折叠栏

1.2 编辑器布局

1.1 节讲解了如何在自定义编辑器中绘制各种类型的编辑器元素，那么如何让这些编辑器元素根据我们的设计进行绘制？接下来介绍如何控制编辑器元素布局的方向、大小及间隙等内容。

1.2.1 水平与垂直布局

如果想要将一些编辑器元素在水平方向上进行排列布局，则需要用到 GUILayout 类或者 EditorGUILayout 类中的 BeginHorizontal()与 EndHorizontal()方法，注意这两种方法一定是成对出现的，否则编辑器布局将会出现异常。相应地，如果想要将这些编辑器元素在垂直方向上进行排列布局，则需要用到 BeginVertical()与 EndVertical()方法。编辑器默认以垂直方向进行布局，也就是说，如果没有调用上述的水平布局或者垂直布局的方法，则编辑器默认会在垂直方向上布局编辑器元素。

在示例脚本中使用垂直布局绘制一组按钮，再用水平布局绘制另一组按钮，查看效果，代码如下：

```
//第1章/ExampleEditor.cs

public override void OnInspectorGUI()
{
    base.OnInspectorGUI();
    Layout();
}
private void Layout()
{
    //默认为垂直布局
    GUILayout.Button("Button1");
    GUILayout.Button("Button2");
    //调用水平布局方法
```

```csharp
GUILayout.BeginHorizontal();
GUILayout.Label("这是一组按钮");
GUILayout.Button("Button3");
GUILayout.Button("Button4");
GUILayout.Button("Button5");
GUILayout.EndHorizontal();
}
```

结果如图1-9所示。

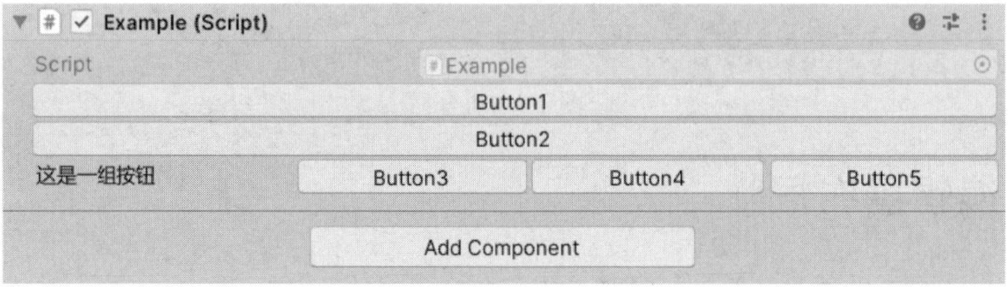

图1-9 水平与垂直布局

布局是支持相互嵌套的，可以在水平布局中嵌套垂直布局，或者在垂直布局中嵌套水平布局。在示例脚本中首先创建一个水平布局，然后在水平布局中嵌套两个垂直布局，在第1个垂直布局中绘制按钮Button1和Button2，在第2个垂直布局中绘制Button3和Button4，代码如下：

```csharp
//第1章/ExampleEditor.cs

public override void OnInspectorGUI()
{
    base.OnInspectorGUI();
    Layout();
}
private void Layout()
{
    //在水平布局中嵌套两个垂直布局
    GUILayout.BeginHorizontal();
    //第1个垂直布局
    GUILayout.BeginVertical();
    GUILayout.Button("Button1");
    GUILayout.Button("Button2");
    GUILayout.EndVertical();
    //第2个垂直布局
    GUILayout.BeginVertical();
    GUILayout.Button("Button3");
    GUILayout.Button("Button4");
    GUILayout.EndVertical();
    GUILayout.EndHorizontal();
```

}
```

结果如图1-10所示。

图1-10 布局嵌套

## 1.2.2 GUI中的滚动列表

如果想要在有限的区域内展示大量的编辑器元素，则需要使用滚动列表，GUILayout和EditorGUILayout类中均有相应的方法，与水平布局和垂直布局一样，BeginScrollView()和EndScrollView()是成对出现的方法。在调用BeginScrollView()方法开启滚动列表时，需要传入一个二维向量类型的参数，该参数表示滚动的位置，也就是滑动条的位置，方法的返回值是滑动条被交互拖动后的值，代码如下：

```
public static Vector2 BeginScrollView(Vector2 scrollPosition, params
GUILayoutOption[] options);
    public static Vector2 BeginScrollView(Vector2 scrollPosition, bool
alwaysShowHorizontal, bool alwaysShowVertical, params GUILayoutOption[]
options);
```

参数alwaysShowHorizontal是指是否始终显示水平方向的滑动条，默认仅在列表中的内容比滚动列表本身的区域更宽时才显示，相应地，alwaysShowHorizontal是指是否始终显示垂直方向的滑动条，默认仅在列表中的内容比滚动列表本身的区域更高时才显示。

自定义一个新的编辑器窗口（如何自定义编辑器窗口将在后面的章节中进行介绍），首先声明一个二维向量类型的变量scrollPosition，然后在窗口中使用BeginScrollView()方法创建一个滚动列表，将scrollPosition作为参数传入并用scrollPosition接收方法的返回值，在滚动列表中绘制一组按钮，代码如下：

```
//第1章/ExampleEditorWindow.cs

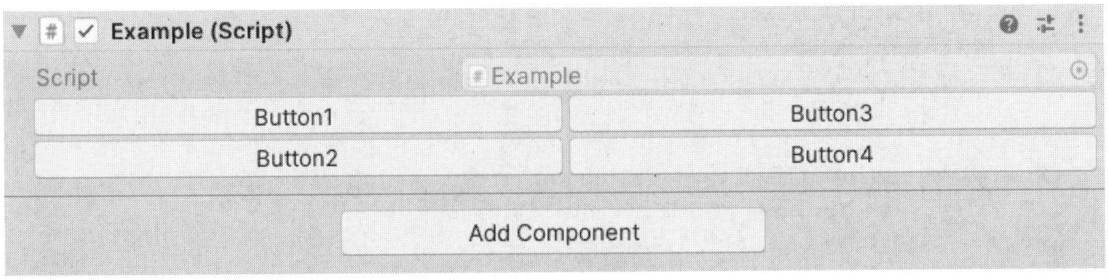

using UnityEngine;
using UnityEditor;

public class ExampleEditorWindow : EditorWindow
{
    [MenuItem("Example/Open Example Editor Window")]
    public static void Open()
    {
```
```

```
        GetWindow<ExampleEditorWindow>().Show();
    }

    private Vector2 scrollPosition;

    private void OnGUI()
    {
        ScrollViewExample();
    }
    private void ScrollViewExample()
    {
        scrollPosition = GUILayout.BeginScrollView(scrollPosition);
        for (int i = 0; i < 50; i++)
        {
            GUILayout.Button("Button" + i);
        }
        GUILayout.EndScrollView();
    }
}
```

结果如图 1-11 所示。

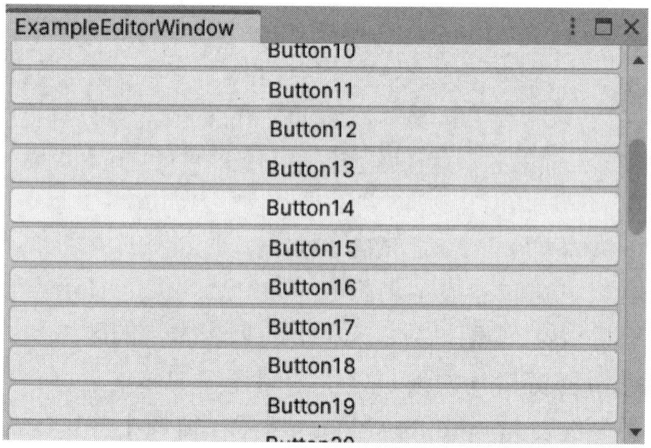

图 1-11 滚动列表

1.2.3 GUI 元素和布局的大小

无论是在绘制编辑器元素的方法中，还是在开始水平布局或垂直布局的方法中都有 GUILayoutOption 的可变参类型参数，表示布局的选项。GUILayout 类中的 Width() 方法用于设定宽度，Height() 方法用于设定高度，它们的返回值均为 GUILayoutOption 类型。

在示例脚本中使用 Button() 方法绘制 3 个按钮，并通过调用 Width() 和 Height() 方法为它们设置不同的大小，代码如下：

```
//第1章/ExampleEditor.cs
```

```
public override void OnInspectorGUI()
{
    base.OnInspectorGUI();
    LayoutOption();
}
private void LayoutOption()
{
    GUILayout.BeginHorizontal();
    GUILayout.Button("Button1",
        GUILayout.Width(50f));
    GUILayout.Button("Button2",
        GUILayout.Width(150f), GUILayout.Height(30f));
    GUILayout.Button("Button3",
        GUILayout.Width(200f), GUILayout.Height(40f));
    GUILayout.EndHorizontal();
}
```

结果如图 1-12 所示。

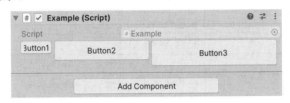

图 1-12　GUI 元素大小

除了 Width() 与 Height() 方法外，GUILayout 类中其他用于控制元素和布局大小的方法见表 1-4。

表 1-4　设定宽和高的方法

方　　法	作　　用	方　　法	作　　用
MaxHeight()	指定最大高度的选项	MinWidth()	指定最小宽度的选项
MaxWidth()	指定最大宽度的选项	ExpandHeight()	指定是否允许垂直扩展的选项
MinHeight()	指定最小高度的选项	ExpandWidth()	指定是否允许水平扩展的选项

在某些自定义编辑器窗口中会使用分割线将窗口划分为多个区域，拖动这些分割线可以灵活地调整各区域的大小。例如，想要将一个窗口划分为左右两个区域，可以创建一个水平布局，在水平布局中创建两个垂直布局，将窗口分为左右两部分，左侧垂直布局使用 Width() 方法设置宽度，在调用该方法时，参数传入的是一个 float 类型的变量而不是一个固定值，右侧垂直布局使用 ExpandWidth() 方法，水平扩展右侧的垂直布局，这样就可以在拖动分割线时通过调整这个变量来调整左右两侧区域的大小。

如何判断分割线被拖曳？需要先了解如何在编辑器中获取用户的输入，此时会用到 Event 类中的 Type 属性，它表示事件的类型，详情见表 1-5。

表 1-5 事件类型

Type	描述
MouseDown	鼠标按下
MouseUp	鼠标抬起
MouseMove	鼠标移动
MouseDrag	鼠标拖动
KeyDown	键盘按键按下
KeyUp	键盘按键抬起
ScrollWheel	鼠标滚轮滚动
Repaint	重绘事件
Layout	布局事件
DragUpdated	拖放操作已更新
DragPerform	拖放操作已执行
DragExited	拖放操作已退出
Ignore	应忽略事件
Used	已经处理了事件
ValidateCommand	验证特殊命令（例如复制和粘贴）
ExecuteCommand	执行特殊命令（例如复制和粘贴）
ContextClick	右击或者在 Mac 上单击了 Control
MouseEnterWindow	鼠标进入某个窗口
MouseLeaveWindow	鼠标离开某个窗口

使用 MouseDown 获取鼠标按下事件，鼠标按下时判断如果是在分割线的区域，则开始拖曳，获取 MouseUp 类型事件后拖曳结束，获取 MouseDrag 类型事件时在拖曳过程中计算鼠标拖动的偏移量，示例代码如下：

```
//第1章/ExampleEditorWindow.cs

using UnityEngine;
using UnityEditor;

public class ExampleEditorWindow : EditorWindow
{
    [MenuItem("Example/Open Example Editor Window")]
    public static void Open()
    {
        GetWindow<ExampleEditorWindow>().Show();
    }

    private const float splitterWidth = 2f;
```

```csharp
private float splitterPos;
private Rect splitterRect;
private bool isDragging;

private void OnEnable()
{
    splitterPos = position.width * .3f;
}

private void OnGUI()
{
    GUILayout.BeginHorizontal();
    {
        GUILayout.BeginVertical(GUILayout.Width(splitterPos));
        GUILayout.Box("左侧区域", GUILayout.ExpandHeight(true),
            GUILayout.ExpandWidth(true));
        GUILayout.EndVertical();

        //分割线，垂直扩展
        GUILayout.Box(string.Empty, GUILayout.Width(splitterWidth),
            GUILayout.ExpandHeight(true));
        //该方法用于获取 GUILayout，最后用于控件的矩形区域
        splitterRect = GUILayoutUtility.GetLastRect();

        GUILayout.BeginVertical(GUILayout.ExpandWidth(true));
        GUILayout.Box("右侧区域", GUILayout.ExpandHeight(true),
            GUILayout.ExpandWidth(true));
        GUILayout.EndVertical();
    }
    GUILayout.EndHorizontal();

    if (Event.current != null)
    {
        //该方法用于在指定区域内显示指定的鼠标光标类型
        EditorGUIUtility.AddCursorRect(splitterRect,
            MouseCursor.ResizeHorizontal);
        switch (Event.current.type)
        {
            //开始拖曳分割线
            case EventType.MouseDown:
                isDragging = splitterRect.Contains(
                    Event.current.mousePosition);
                break;
            case EventType.MouseDrag:
                if (isDragging)
                {
                    splitterPos += Event.current.delta.x;
```

```
                //限制其最大值和最小值
                splitterPos = Mathf.Clamp(splitterPos,
                    position.width * .2f, position.width * .8f);
                Repaint();
            }
            break;
        //结束拖曳分割线
        case EventType.MouseUp:
            if (isDragging)
                isDragging = false;
            break;
    }
}
```

结果如图 1-13 所示。

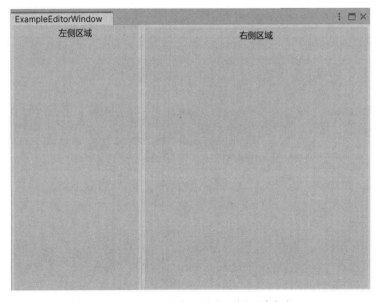

图 1-13　拖曳调整窗口左右两侧区域大小

1.2.4　GUI 元素之间的间隙

编辑器布局的间隙可以通过 GUILayout 类中的 Space() 和 FlexibleSpace() 方法进行设定，两者的区别在于，前者通过一个 float 类型的参数设定间隙值，而后者是一个无参的方法，根据字面意思去理解，它会根据编辑器窗口的大小灵活地调整间隙的大小，也就是插入灵活的空白元素。

在示例脚本中创建两个水平布局，均绘制两个按钮 Button1 和 Button2，在第 1 个水平布局的两个按钮之间使用 Space() 方法设置固定的间隔（50 像素），在第 2 个水平布局的两个按钮之间使用 FlexibleSpace() 方法插入灵活的空白元素，代码如下：

```
//第1章/ExampleEditor.cs

public override void OnInspectorGUI()
{
    base.OnInspectorGUI();
    Space();
}
private void Space()
{
    GUILayout.BeginHorizontal();
    GUILayout.Button("Button1", GUILayout.Width(80f));
    //固定间隔（50像素）
    GUILayout.Space(50f);
    GUILayout.Button("Button2", GUILayout.Width(80f));
    GUILayout.EndHorizontal();

    GUILayout.BeginHorizontal();
    GUILayout.Button("Button1", GUILayout.Width(80f));
    //灵活调整间隙（Button1在最左侧，Button2在最右侧）
    GUILayout.FlexibleSpace();
    GUILayout.Button("Button2", GUILayout.Width(80f));
    GUILayout.EndHorizontal();
}
```

结果如图 1-14 所示。

图 1-14　元素间隙

第 2 章 自定义功能菜单

本章将介绍编辑器开发中常用的 3 个特性：MenuItem、ContextMenu 和 ContextMenuItem。这些特性可以帮助开发者更好地管理和组织编辑器工具菜单。本章将详细讲解这些特性的使用方法和示例，以帮助读者掌握它们的用法。

2.1 MenuItemAttribute

MenuItem 特性的作用是创建新的功能菜单项，涉及 Unity 顶部的菜单、在 Hierarchy 窗口右击后弹出的菜单、在 Project 窗口右击后弹出的菜单及 Inspector 窗口中组件的下拉列表菜单，该特性用于静态方法，其构造函数的代码如下：

```
public MenuItem (string itemName);
public MenuItem (string itemName, bool isValidateFunction);
public MenuItem (string itemName, bool isValidateFunction, int priority);
```

参数 itemName 表示菜单项的路径，例如创建 UGUI 中文本组件的功能，其菜单项路径为 GameObject/UI/Text。参数 isValidateFunction 表示方法是否为验证方法，默认值为 false，验证方法将在调用具有相同菜单项路径的方法之前被调用，用于验证这个具有相同菜单项路径的方法是否可被调用。参数 priority 表示优先级，它决定了菜单项的显示顺序，默认值为 1000，数值越小越靠上显示，若相邻功能菜单项之间相差大于 10，则会被分割线分割。

2.1.1 自定义 Unity 顶部的功能菜单

如图 2-1 所示，在编辑器顶部默认有 File、Edit、Assets、GameObject 等菜单栏，可以通过 MenuItem 增加新的菜单栏，也可以在已有的菜单栏中增加新的菜单项。

图 2-1 Unity 顶部默认的功能菜单

例如新增两个功能，菜单项路径分别是 Window/Function1、Example/Function2，Function1

在默认已有的 Window 菜单栏中,而默认没有 Example 这个菜单栏,因此自定义后在编辑器顶部会出现一个新的菜单栏 Example,Function2 在其中,代码如下:

```
//第2章/MenuItemExample.cs

using UnityEngine;
using UnityEditor;

public class MenuItemExample
{
    [MenuItem("Window/Function1")]
    static void Function1()
    {
        Debug.Log("Function1 Invoke.");
    }
    [MenuItem("Example/Function2")]
    static void Function2()
    {
        Debug.Log("Function2 Invoke.");
    }
}
```

结果如图 2-2 所示。

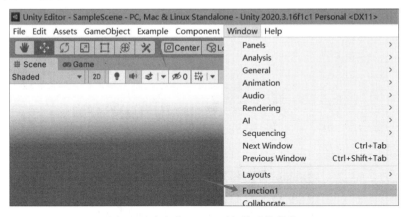

图 2-2　自定义 Unity 顶部的功能菜单

参数 itemName 除了可以设置菜单项的路径之外,还可以设置执行该菜单项功能的快捷键,快捷键与设置快捷键所需的字符串的对应关系见表 2-1。

表 2-1　快捷键与设置快捷键所需的字符串的对应关系

快　捷　键	字　符　串
Ctrl on Windows / Commond on OSX	%
Shift	#

续表

快 捷 键	字 符 串
Alt	&
Arrow Keys	LEFT/RIGHT/UP/DOWN
F Keys	F1…F12
Home、End、Page Up、Page Down	HOME、END、PGUP、PGDN

例如想要使用快捷键 Shift+E 时执行 Function1，那么菜单路径需要设置为"Window/Function1 #E"，想要使用快捷键 Alt+R 时执行 Function2，那么菜单路径需要设置为"Example/Function2 &R"。还可以设置组合快捷键，例如 Shift+Alt+E，快捷键字符串需要设置为"#&E"，注意菜单项路径与设置快捷键的字符串之间包含一个空格，示例代码如下：

```csharp
//第 2 章/MenuItemExample.cs

[MenuItem("Window/Function1 #E")]
private static void Function1()
{
    Debug.Log("Function1 Invoke.");
}
[MenuItem("Example/Function2 &R")]
private static void Function2()
{
    Debug.Log("Function2 Invoke.");
}
[MenuItem("Example/Function3 #&E")]
private static void Function3()
{
    Debug.Log("Function3 Invoke.");
}
```

结果如图 2-3 所示。

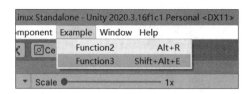

图 2-3　设置功能快捷键

当参数 isValidateFunction 为 true 时表明这是一个验证方法，验证方法的返回值为布尔类型，当方法的返回值为 true 时，该菜单项才可以使用，否则置灰不可单击。

例如创建一个新的功能菜单项，该功能执行时打印当前所选中的游戏物体的名称，这种情况便可以使用验证方法来判断当前是否选中了某个游戏物体，如果没有选中，则该菜单项无法单击使用，代码如下：

```
//第 2 章/MenuItemExample.cs

[MenuItem("GameObject/LogName", priority = 0)]
private static void LogName()
{
    Debug.Log(Selection.activeGameObject.name);
}
[MenuItem("GameObject/LogName", true)]
private static bool LogNameValidate()
{
    return Selection.activeGameObject != null;
}
```

如图 2-4 所示，在 Hierarchy 窗口或 Project 窗口没有选中任何游戏物体时，Selection 类中的静态变量 activeGameObject 为 null，此时该菜单项为置灰状态。

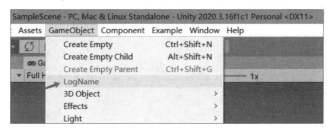

图 2-4 菜单项置灰

在同一个菜单栏中增加 3 个功能 Function4、Function5 与 Function6，将 priority 分别设为 20、10 和 31。因为数值越小越靠上显示，所以它们的显示顺序依次是 Function5、Function4、Function6，而且由于 Function4 与 Function6 之间的数值相差大于 10，因此它们之间被分割线分割，代码如下：

```
//第 2 章/MenuItemExample.cs

[MenuItem("Example/Function4", false, 20)]
static void Function4() { }
[MenuItem("Example/Function5", false, 10)]
static void Function5() { }
[MenuItem("Example/Function6", false, 31)]
static void Function6() { }
```

结果如图 2-5 所示。

1. 网格合并

本节介绍一个网格合并的功能案例，网格合并是优化中常用的小手段，其目的是减少 drawcall，大量的 drawcall 会造成 CPU 的性能瓶颈。如图 2-6 所示，船只里的钢材货物由诸多钢材模型堆砌而成，将其放在一个空场景中查看 Stats 面板中的统计数据发现 Batches 指数

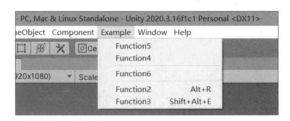

图 2-5 设置优先级

为 94。通过 MenuItem 创建一个新的功能菜单项，该功能执行时会将所有选中的游戏物体的 MeshFilter 组件中的网格合并。使用该功能对这些钢材模型的网格进行合并后，Batches 指数由 94 变为 42。

图 2-6 钢材

实现网格合并功能需要用到 Mesh 类中的 CombineMeshes()方法，代码如下：

```
public void CombineMeshes(CombineInstance[] combine, bool mergeSubMeshes);
```

参数 combine 表示要合并的网格实例，mergeSubMeshes 表示是否合并到单个子网格中，如果所有网格共享相同的材质，则应将其设置为 true。

要合并的网格需要在所选中的游戏物体中获取，mergeSubMeshes 参数值通过弹出一个对话框，让用户选择是否合并到单个子网格中，然后遍历当前所选中的物体，获取物体上 MeshFilter 组件中的网格并存储到一个列表中。当前所选中的物体可以在 Selection 类的静态变量中获取，见表 2-2。

表 2-2 Selection 类的静态变量

静态变量	描述
activeContext	当前的上下文对象，与 SetActiveObjectWithContext 方法搭配使用

续表

静态变量	描述
activeGameObject	当前在 Hierarchy 或 Project 窗口中所选中的游戏物体
activeInstanceID	当前在 Hierarchy 或 Project 窗口中所选中的对象的实例 ID
activeObject	当前在 Hierarchy 或 Project 窗口中所选中的对象
activeTransform	当前在 Hierarchy 窗口中所选中的 Transform
assetGUIDs	当前在 Project 窗口中所选中的资源的 guid 集合
gameObjects	当前在 Hierarchy 或 Project 窗口中所选中的游戏物体集合
instanceIDs	当前在 Hierarchy 或 Project 窗口中所选中的对象的实例 ID 集合
objects	当前在 Hierarchy 或 Project 窗口中所选中的对象集合
transforms	当前在 Hierarchy 窗口中所选中的 Transform 集合

将所选中物体上的 MeshRenderer 组件放到一个集合中，如果 mergeSubMeshes 参数为 true，则可获取第 1 个 MeshRenderer 组件的材质，否则获取所有 MeshRenderer 组件的材质。创建一个新的物体为其添加 MeshFilter 组件和 MeshRenderer 组件，创建一个新的网格，调用网格合并的方法，将合并后的网格赋值给 MeshFilter 组件，将材质赋值给 MeshRenderer 组件。

通过 MenuItem 创建网格合并功能的菜单项，为该功能设置一个快捷键 Ctrl+M，并添加一个验证方法，只有当前所选中的游戏物体数量是两个或两个以上时才可以进行网格合并，因此验证方法中应返回 Selection.gameObjects 集合的长度是否大于或等于 2，代码如下：

```
//第2章/MeshCombiner.cs

using UnityEditor;
using UnityEngine;
using System.Linq;
using System.Collections.Generic;

//<summary>
//Mesh 网格合并
//</summary>
public class MeshCombiner
{
    [MenuItem("Example/Mesh/Combine %M")]
    public static void Execute()
    {
        //弹出对话框，让用户选择是否合并到单个子网格中
        bool mergeSubMeshes = EditorUtility.DisplayDialog(
            "网格合并", "是否合并到单个子网格中", "是", "否");
        //列表存储选中的所有网格实例
        List<CombineInstance> instances = new List<CombineInstance>();
        //遍历当前所选中的物体
```

```csharp
for (int i = 0; i < Selection.gameObjects.Length; i++)
{
    //非空判断
    GameObject go = Selection.gameObjects[i];
    if (go == null) continue;
    MeshFilter meshFilter = go.GetComponent<MeshFilter>();
    if (meshFilter == null) continue;
    Mesh target = meshFilter.sharedMesh;
    if (target == null) continue;
    //添加到列表中
    instances.Add(new CombineInstance
    {
        mesh = target,
        transform = meshFilter.transform.localToWorldMatrix
    });
}
//获取所有选中的MeshRenderer组件
var mrs = Selection.gameObjects.Select(
    m => m.GetComponent<MeshRenderer>()).ToArray();
//如果合并到单个子网格中，则可获取第1个MeshRenderer组件的材质
//否则获取所有MeshRenderer组件的材质
var materials = mergeSubMeshes
    ? mrs.First().sharedMaterials.ToArray()
    : mrs.SelectMany(m => m.sharedMaterials).ToArray();
//创建一个新的物体
GameObject instance = new GameObject("New Mesh Combined");
//为其添加MeshFilter组件
MeshFilter filter = instance.AddComponent<MeshFilter>();
//创建新的网格
filter.mesh = new Mesh { name = instance.name };
//网格合并
filter.sharedMesh.CombineMeshes(instances.ToArray(),
    mergeSubMeshes);
//为新物体添加MeshRenderer组件
MeshRenderer renderer = instance.AddComponent<MeshRenderer>();
//为MeshRenderer组件设置材质
renderer.sharedMaterials = materials.ToArray();
}

[MenuItem("Example/Mesh/Combine %M", true)]
public static bool Validate()
{
    //只有当前选中的游戏物体至少为两个时才可以进行网格合并
    return Selection.gameObjects.Length >= 2;
}
}
```

网格合并的功能菜单项如图 2-7 所示。

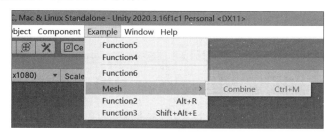

图 2-7　网格合并的功能菜单项

2. 网格提取

如果想要把合并后的网格作为资产进行保存，则可以使用 AssetDatabase 类中的 CreateAsset()方法创建这个网格资产，调用该方法需要在参数中指定保存资产的路径，这个路径以 Assets/开头，也就是相对于项目文件夹的路径，并且路径中需要包含文件的扩展名。各类型资产对应的文件扩展名见表 2-3。

表 2-3　各类型资产对应的文件扩展名

资 产 类 型	文件扩展名	资 产 类 型	文件扩展名
材质	.mat	动画	.anim
立方体贴图	.cubemap	其他	.asset
GUI 皮肤	.GUISkin		

通过 MenuItem 创建网格提取功能的菜单项，当功能执行时，在当前所选中的物体上获取 MeshFilter 组件，实例化组件中的网格，然后创建这个网格资产，代码如下：

```
//第 2 章/MeshExtracter.cs

using System;
using UnityEditor;
using UnityEngine;

//<summary>
//Mesh 网格提取器
//</summary>
public class MeshExtracter
{
    [MenuItem("Example/Mesh/Extract")]
    public static void Execute()
    {
        //非空判断
        MeshFilter meshFilter = Selection.activeGameObject
            .GetComponent<MeshFilter>();
        if (meshFilter == null) return;
```

```
            Mesh mesh = meshFilter.sharedMesh;
            if (mesh == null) return;
            try
            {
                Mesh instance = UnityEngine.Object.Instantiate(mesh);//实例化
                AssetDatabase.CreateAsset(instance,
                    string.Format("Assets/{0}.asset", mesh.name));       //创建资产
                AssetDatabase.Refresh();                                 //刷新
                Selection.activeObject = instance;                       //选中
            }
            catch (Exception error)
            {
                Debug.Log(string.Format("{0}提取Mesh网格失败: {1}",
                    Selection.activeGameObject.name, error));
            }
        }


        [MenuItem("Example/Mesh/Extract", true)]
        public static bool Validate()
        {
            return Selection.activeGameObject != null;
        }
    }
}
```

网格提取的功能菜单项如图2-8所示。

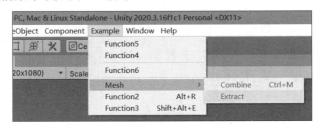

图2-8　网格提取的功能菜单项

2.1.2　自定义Hierarchy窗口右键功能菜单

当菜单项路径以GameObject/开头时，该菜单项可以被添加到Hierarchy层级窗口中右击后弹出的菜单中。需要注意的是，当在这个菜单中添加菜单项时，优先级设置的数值需要小于50，如果大于或等于50，则该菜单项将不会显示在其中。

例如，在示例脚本中创建7个菜单项，将路径设置为GameObject/GoFunction，以序号1~7结尾，为它们设置不同的优先级，其中GoFunction6、GoFunction7的优先级分别为50、100，代码如下：

```
//第2章/MenuItemExample.cs
```

```
[MenuItem("GameObject/GoFunction1", false, -1000)]
static void GoFunction1(){ }
[MenuItem("GameObject/GoFunction2", false, -100)]
static void GoFunction2(){ }
[MenuItem("GameObject/GoFunction3", false, 0)]
static void GoFunction3(){ }
[MenuItem("GameObject/GoFunction4", false, 10)]
static void GoFunction4(){ }
[MenuItem("GameObject/GoFunction5", false, 49)]
static void GoFunction5(){ }
[MenuItem("GameObject/GoFunction6", false, 50)]
static void GoFunction6(){ }
[MenuItem("GameObject/GoFunction7", false, 100)]
static void GoFunction7(){ }
```

结果如图 2-9 所示，可以发现菜单中没有 GoFunction6、GoFunction7 两项。

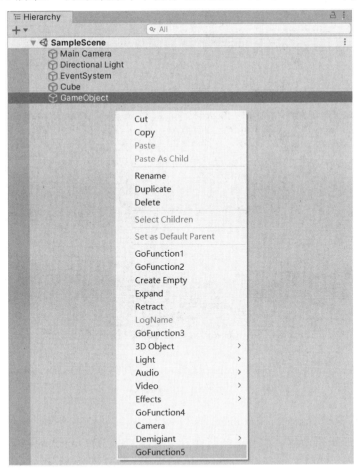

图 2-9　自定义 Hierarchy 窗口右键功能菜单

1. 创建阶梯物体

2.1.1 节实现了网格合并与提取功能，现在创建 5 个 Cube 物体，通过设置不同的坐标和缩放值将它们拼接为一个阶梯形状，并使用网格合并功能将它们合并为一个物体，然后使用网格提取功能提取阶梯网格，将其作为资产进行保存，如图 2-10 所示。

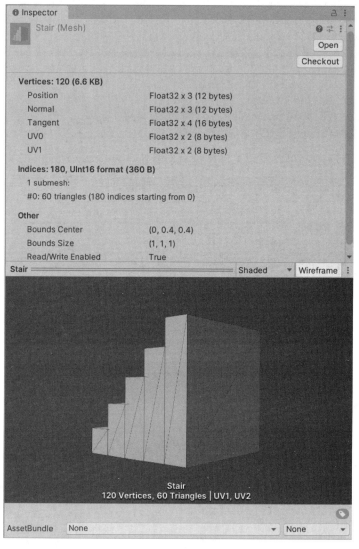

图 2-10 阶梯网格

在 Hierarchy 窗口右击时，3D Object 的菜单栏中可以创建立方体、球体、胶囊体、圆柱体，以及面片等形状的物体。本节添加这样一个功能，在 3D Object 的菜单中增加一个 Stair 的选项，当执行该功能时在场景中创建一个新的物体，为其添加 MeshFilter 和 MeshRenderer 组件，Mesh 使用如图 2-10 所示的阶梯网格，Material 使用默认的材质球，代码如下：

```
//第2章/CreateStair.cs

using UnityEngine;
using UnityEditor;

public class CreateStair
{
    //<summary>
    //创建 3D 物体（阶梯）
    //</summary>
    [MenuItem("GameObject/3D Object/Stair")]
    public static void Execute()
    {
        GameObject go = new GameObject("Stair");
        MeshFilter mf = go.AddComponent<MeshFilter>();
        MeshRenderer mr = go.AddComponent<MeshRenderer>();
        //加载阶梯网格资产
        Mesh stairMesh = AssetDatabase.LoadAssetAtPath<Mesh>(
            "Assets/Mesh/Stair.asset");
        mf.sharedMesh = stairMesh;
        mr.sharedMaterial = new Material(Shader.Find("Standard"));
        //如果当前在 Hierarchy 窗口中已经选中了某个游戏物体
        //将新创建的阶梯物体设为它的子级
        if (Selection.activeTransform != null)
        {
            go.transform.SetParent(Selection.activeTransform);
            go.transform.localPosition = Vector3.zero;
            go.transform.localRotation = Quaternion.identity;
            go.transform.localScale = Vector3.one;
        }
        Selection.activeTransform = go.transform;
    }
}
```

创建阶梯的功能菜单项，如图 2-11 所示。

2. 展开或收起游戏物体层级

在 Hierarchy 窗口中选中某个游戏物体后，通过快捷键 Alt+RightArrow 可以展开该物体的层级，通过快捷键 Alt+LeftArrow 可以收起该物体的层级，本节通过 MenuItem 将这两个功能添加到当右击层级窗口时弹出的菜单中。

展开或收起物体层级的方法在 UnityEditor 的内部类 SceneHierarchyWindow 中，通过反编译工具查看其源码，可以看到方法名为 SetExpandedRecursive，如图 2-12 所示。

参数 id 表示游戏物体的实例 ID，当 expand 为 true 时表示展开物体层级，当 expand 为 false 时表示收起物体层级，想要调用该方法需要使用反射的方式，代码如下：

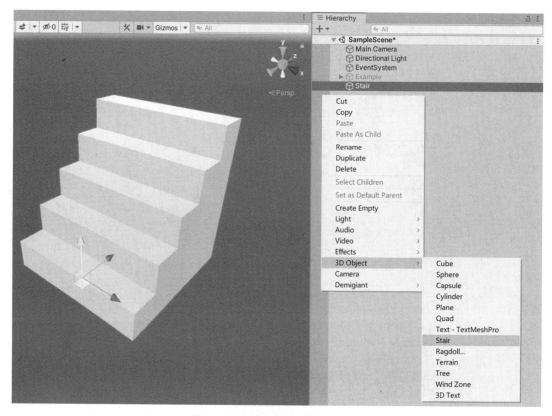

图 2-11 创建阶梯的功能菜单项

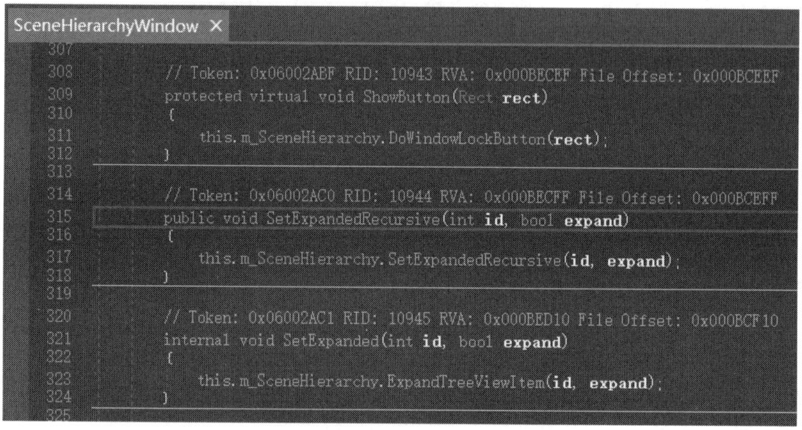

图 2-12 SetExpandedRecursive()

```
//第 2 章/HierarchyWindowUtility.cs

using System;
using UnityEditor;
```

```csharp
using System.Reflection;

public class HierarchyWindowUtility
{
    [MenuItem("GameObject/Expand", priority = 0)]
    public static void Expand()
    {
        //展开物体层级
        SetExpandedRecursive(true);
    }
    [MenuItem("GameObject/Retract", priority = 0)]
    public static void Retract()
    {
        //收起物体层级
        SetExpandedRecursive(false);
    }
    private static void SetExpandedRecursive(bool expand)
    {
        //获取 SceneHierarchyWindow 类型
        Type type = typeof(EditorWindow).Assembly.GetType(
            "UnityEditor.SceneHierarchyWindow");
        //获取 SetExpandedRecursive 方法
        MethodInfo methodInfo = type.GetMethod("SetExpandedRecursive",
            BindingFlags.Public | BindingFlags.Instance);
        //根据类型获取窗口实例
        EditorWindow window = EditorWindow.GetWindow(type);
        object[] array = new object[2];
        array[1] = expand;
        for (int i = 0; i < Selection.transforms.Length; i++)
        {
            array[0] = Selection.transforms[i].gameObject.GetInstanceID();
            //以反射方式调用 SetExpandedRecursive 方法
            methodInfo.Invoke(window, array);
        }
    }
}
```

如图 2-13 所示，当在 Hierarchy 窗口选中某个游戏物体时，可以通过右击后弹出的菜单中的 Expand 和 Retract 功能实现物体层级的展开和收起。

2.1.3　自定义 Project 窗口右键功能菜单

当菜单项路径以 Assets/开头时，该菜单项可以被添加到当右击 Project 工程目录窗口时弹出的菜单中，不同于在 Hierarchy 窗口右击后弹出的菜单中增加菜单项的是，此处增加的菜单项没有优先级取值的限制，示例代码如下：

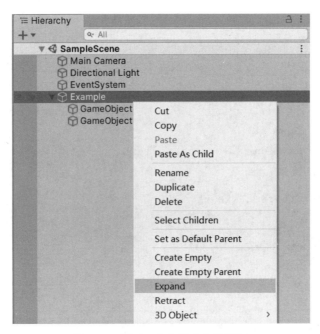

图 2-13　展开、收起游戏物体层级

```
//第 2 章/MenuItemExample.cs

[MenuItem("Assets/Function7")]
private static void Function7()
{
    Debug.Log("Function7 Invoke.");
}
[MenuItem("Assets/Function8")]
private static void Function8()
{
    Debug.Log("Function8 Invoke.");
}
[MenuItem("Assets/Function9")]
private static void Function9()
{
    Debug.Log("Function9 Invoke.");
}
```

结果如图 2-14 所示。

在 Project 窗口右击时，弹出的菜单中包含一个 Select Dependencies 功能，它用于获取指定资产的依赖项，该功能可以通过 AssetDatabase 类中的 GetDependencies()方法实现。但是在某些情况下，开发者想要知道某个资产被其他哪些资产引用了，而 AssetDatabase 类中没有提供获取这些引用项的方法。因此，本节通过 MenuItem 在 Project 窗口被右击时弹出的菜单中新增 1 个 Select Reference 功能。

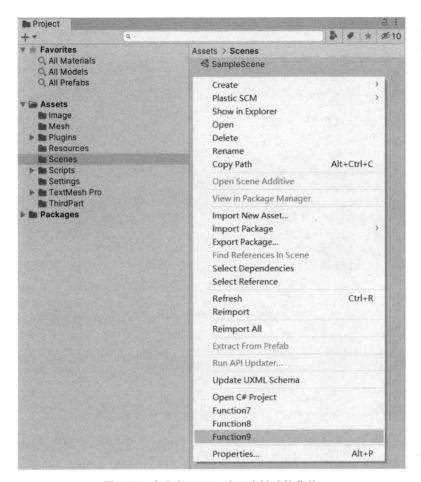

图 2-14　自定义 Project 窗口右键功能菜单

当执行该功能时，首先获取项目中所有资产的路径，遍历这些资产路径获取每项资产的依赖项，并将该资产的路径及对应的依赖项集合存储到一个字典中，然后遍历这个字典并查询资产的依赖项集合，如果依赖项集合包含当前选中的资产，则该资产便是当前选中资产的引用项，代码如下：

```
//第 2 章/AssetReference.cs

using UnityEngine;
using UnityEditor;
using System.Linq;
using System.Collections.Generic;

public class AssetReference
{
    [MenuItem("Assets/Select Reference", priority = 30)]
    public static void GetReference()
```

```csharp
{
    //字典存放资产的依赖关系
    Dictionary<string, string[]> map =
        new Dictionary<string, string[]>();
    //获取所有资产的路径
    string[] paths = AssetDatabase.GetAllAssetPaths();
    //遍历并建立资产间的依赖关系
    for (int i = 0; i < paths.Length; i++)
    {
        string path = paths[i];
        //根据资产路径获取该资产的依赖项
        var dependencies = AssetDatabase.GetDependencies(path).ToList();
        //获取依赖项时会包含该资产本身,将资产本身移除
        dependencies.Remove(path);
        //加入字典
        if (dependencies.Count > 0)
            map.Add(path, dependencies.ToArray());
        //进度条
        EditorUtility.DisplayProgressBar("获取资产依赖关系",
            path, (float)i + 1 / paths.Length);
    }
    EditorUtility.ClearProgressBar();

    //当前所选中资产的路径
    string assetPath = AssetDatabase
        .GetAssetPath(Selection.activeObject);
    //所有引用项的资产路径
    string[] reference = map.Where(
        m => m.Value.Contains(assetPath)).Select(m => m.Key).ToArray();
    //根据路径加载引用项资产并存储到列表中
    List<Object> objects = new List<Object>();
    for (int i = 0; i < reference.Length; i++)
        objects.Add(AssetDatabase.LoadMainAssetAtPath(reference[i]));
    //选中所有的引用项
    Selection.objects = objects.ToArray();
}

[MenuItem("Assets/Select Reference", true)]
public static bool GetReferencesValidate()
{
    return Selection.activeObject != null;
}
}
```

结果如图 2-15 所示。

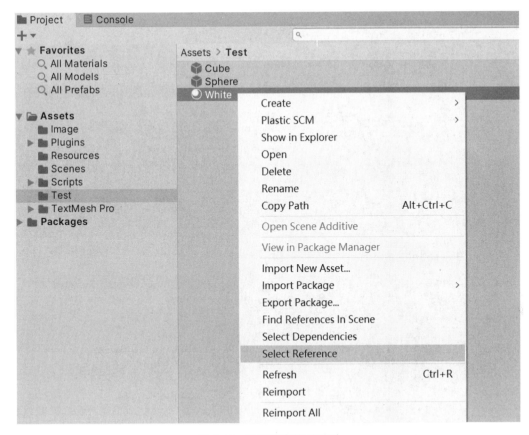

图 2-15　获取资产的引用项

2.1.4　自定义组件下拉列表功能菜单

MenuItem 还可以为组件添加下拉列表功能菜单项，菜单路径需要以 "CONTEXT/组件类名/" 开头。例如，想要为 Image 组件添加一个功能 Convert2RawImage，其目的是使用 RawImage 组件替换 Image 组件，那么 itemName 需要设置为 "CONTEXT/Image/ Convert2RawImage"，代码如下：

```
//第2章/Image2RawImage.cs

using UnityEditor;
using UnityEngine;
using UnityEngine.UI;

public class Image2RawImage
{
    [MenuItem("CONTEXT/Image/Convert2RawImage")]
    public static void Execute()
    {
```

```csharp
        GameObject selected = Selection.activeGameObject;
        //获取选中物体的 Image 组件
        Image image = selected.GetComponent<Image>();
        //缓存 Sprite
        Sprite sprite = image.sprite;
        //销毁 Image 组件
        Object.DestroyImmediate(image);
        //添加 RawImage 组件
        RawImage rawImage = selected.AddComponent<RawImage>();
        //如果 Sprite 不为 null，则将其 texture 赋值给 RawImage 的 texture
        if (sprite != null)
            rawImage.texture = sprite.texture;
    }
}
```

结果如图 2-16 所示。

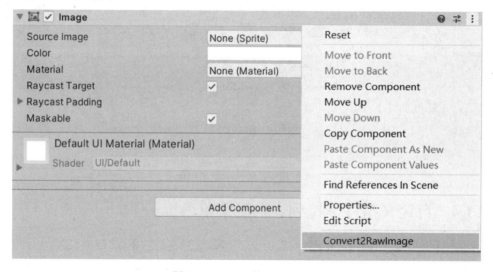

图 2-16 Image 转 RawImage

本节通过 MenuItem 为 BoxCollider 组件添加一个名为 Encapsulate 的功能菜单项，在执行该功能时，BoxCollider 可以自动适配大小，不仅让边界盒（Bounds）包裹住这个物体本身，还要包裹住它的子物体。

实现该功能需要用到 Bounds 类中的 Encapsulate()方法，它用于扩展大小以包裹其他的边界盒或点。该功能菜单项需要被添加到 BoxCollider 组件的下拉列表中，因此 MenuItem 的菜单路径以 "CONTEXT/BoxCollider/" 开头。

在方法中获取当前选中的 BoxCollider 组件下的所有 MeshRenderer 组件，创建一个新的边界盒，调用 Encapsulate()方法来包裹这些 MeshRenderer 中的边界盒，最终将该边界盒的中心点和大小赋值给 BoxCollider 组件，代码如下：

```csharp
//第2章/AutoBoxCollider.cs

using UnityEditor;
using UnityEngine;

public class AutoBoxCollider
{
    [MenuItem("CONTEXT/BoxCollider/Encapsulate")]
    public static void Execute()
    {
        Transform selected = Selection.activeTransform;
        var boxCollider = selected.GetComponent<BoxCollider>();
        //记录坐标、旋转值、缩放值
        Vector3 position = selected.position;
        Quaternion rotation = selected.rotation;
        Vector3 scale = selected.localScale;
        //坐标旋转归零
        selected.position = Vector3.zero;
        selected.rotation = Quaternion.identity;
        //创建边界盒并计算
        Bounds bounds = new Bounds(Vector3.zero, Vector3.zero);
        var renders = selected
            .GetComponentsInChildren<MeshRenderer>(true);
        for (int j = 0; j < renders.Length; j++)
        {
            bounds.Encapsulate(renders[j].bounds);
        }
        boxCollider.center = bounds.center;
        Vector3 size = bounds.size;
        size.x /= scale.x;
        size.y /= scale.y;
        size.z /= scale.z;
        boxCollider.size = size;
        //恢复坐标旋转
        selected.position = position;
        selected.rotation = rotation;
    }
}
```

如图 2-17 所示,在场景中选中一个立方体,其他两个立方体是它的子物体,执行 Encapsulate 功能后,选中的立方体的 BoxCollider 组件会进行大小适配,以便包裹其他两个立方体。

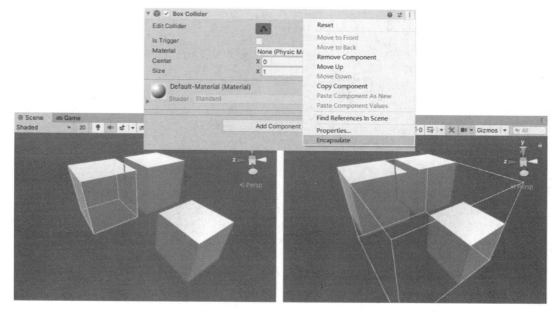

图 2-17 Box Collider 自适配大小

2.2 ContextMenuAttribute

ContextMenu 特性同样可以在组件的下拉列表菜单中创建新的功能菜单项，构造函数的代码如下：

```
public ContextMenu (string itemName);
public ContextMenu (string itemName, bool isValidateFunction);
public ContextMenu (string itemName, bool isValidateFunction, int priority);
```

可以看出，ContextMenu 特性与 2.1 节中 MenuItem 特性的构造函数参数是一致的，它们的使用方法也基本一致，不同的是 MenuItem 用于静态方法，可以为 Unity 默认已有的组件添加下拉列表功能菜单项，例如 Image/Convert2RawImage，而 ContextMenu 在自定义组件中用于非静态方法，可以为该自定义组件添加下拉列表功能菜单项。

以自定义组件 ContextMenuExample 为例，在组件中添加一个 Function1 方法，为该方法添加 ContextMenu 特性，代码如下：

```
//第2章/ContextMenuExample.cs

using UnityEngine;

public class ContextMenuExample : MonoBehaviour
{
    [ContextMenu("Function1")]
    private void Function1()
```

```
    {
        Debug.Log("TODO Something.");
    }
}
```

结果如图2-18所示。

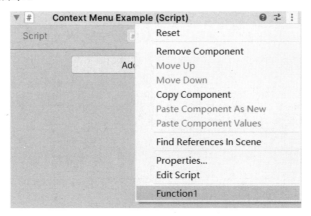

图2-18　为自定义组件添加下拉列表功能菜单项

2.3　ContextMenuItemAttribute

ContextMenuItem 特性的构造函数的代码如下：

```
public ContextMenuItemAttribute (string name, string function);
```

参数 name 表示菜单项的名称，function 表示使用该功能要调用的方法名，该特性用于组件中的字段，为字段添加该特性后，在检视面板右击该字段时会弹出对应的功能菜单项。

例如，在示例组件 ContextMenuItemExample 中声明一个整数类型变量 size，为其添加 ContextMenuItem 特性，其目的是在使用该特性对应的功能时将 size 重置为 0，代码如下：

```
//第2章/ContextMenuItemExample.cs

using UnityEngine;

public class ContextMenuItemExample : MonoBehaviour
{
    [ContextMenuItem("Reset", "ResetSize")]
    public int size = 0;

    void ResetSize()
    {
        size = 0;
    }
}
```

结果如图 2-19 所示。

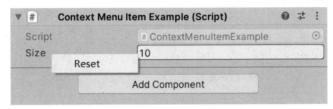

图 2-19　为字段添加右键功能菜单项

第 3 章 自定义检视面板

本章将介绍如何使用 CustomEditor 特性和 Editor 类自定义组件的检视面板,以及如何使用 CustomPropertyDrawer 特性和 PropertyDrawer 类自定义属性在检视面板中的绘制。为了帮助读者更好地理解这些知识,在本章中提供了详细的示例。

3.1 创建自定义编辑器类

如果想要自定义一个组件的检视面板(Inspector)如何绘制,则需要为组件创建自定义编辑器类,该类继承了 Editor 类,并为其添加了 CustomEditor 特性,特性的构造函数的代码如下:

```
public CustomEditor (Type inspectedType);
public CustomEditor (Type inspectedType, bool editorForChildClasses);
```

参数 inspectedType 表示检视的类型,即自定义哪种类型组件的检视面板,editorForChildClasses 表示是否为其派生类也使用同样的检视面板,默认值为 false。

以 CustomComponent 组件为例,为其创建自定义编辑器类 CustomComponentEditor,代码如下:

```
using UnityEngine;
using UnityEditor;

[CustomEditor(typeof(CustomComponent))]
public class CustomComponentEditor : Editor { }
```

类似于 MonoBehaviour 的 Awake()、OnEnable()、OnDisable()、OnDestroy()等生命周期函数,Editor 类中也有相应的回调方法,见表 3-1。

表 3-1 Editor 类中的回调方法

方　　法	详　　解
Awake()	当组件所挂载的物体被选中时,该函数会被调用
OnEnable()	当组件所挂载的物体被选中时,该函数会被调用,晚于 Awake()执行

方法	详解
OnValidate()	当组件的检视面板的值被修改时,该函数会被调用
OnDisable()	当组件所挂载的物体被取消选中时,该函数会被调用
OnDestroy()	当组件所挂载的物体被取消选中时,该函数会被调用,晚于 OnDisable()执行

其中,Awake()、OnEnable()方法用于进行初始化操作,例如,如果想要获取所检视的组件对象,则可以在 OnEnable()中将 Editor 中的 target 对象转换为目标类型,target 表示所检视的对象,类型为 Object,示例代码如下:

```
//第3章/CustomComponentEditor.cs

using UnityEngine;
using UnityEditor;

[CustomEditor(typeof(CustomComponent))]
public class CustomComponentEditor : Editor
{
    private CustomComponent component;

    private void OnEnable()
    {
        component = target as CustomComponent;
    }
}
```

3.1.1 如何自定义检视面板中的 GUI 内容

Editor 类中的虚方法 OnInspectorGUI()定义了组件的检视面板如何绘制,因此重写该方法即可实现自定义。如果想要扩展检视面板,则可以保留 base.OnInspectorGUI()的调用,再添加扩展的内容,也可以不保留,完全自定义。

首先给 CustomComponent 组件添加一些不同类型的变量,再来展示如何在 OnInspectorGUI()方法中添加控件来编辑这些变量的值,代码如下:

```
//第3章/CustomComponent.cs

using UnityEngine;

public class CustomComponent : MonoBehaviour
{
    public int intValue;
    [SerializeField] private string stringValue;
    [SerializeField] private bool boolValue;
    [SerializeField] private GameObject go;
```

```
    public enum ExampleEnum
    {
        Enum1,
        Enum2,
        Enum3,
    }
    [SerializeField] private ExampleEnum enumValue;
}
```

因为整数类型的字段 intValue 使用了 public 进行修饰，所以在编辑器类中可以直接访问和修改其值。那么其他使用 private 修饰的私有类型的字段，应该如何在编辑器类中访问和修改它们的值呢？

有两种方式，一种是使用 SerializedProperty 序列化属性；另一种是使用反射，一般推荐使用前者。

序列化属性通过 Editor 中的 serializedObject 属性调用 FindProperty() 方法获取，该属性指的是当前检视的序列化对象，当调用 FindProperty() 方法时将要获取的序列化属性的字段名称作为参数传入即可，代码如下：

```
//第3章/CustomComponentEditor.cs

using UnityEngine;
using UnityEditor;
using System.Reflection;

[CustomEditor(typeof(CustomComponent))]
public class CustomComponentEditor : Editor
{
    private CustomComponent component;
    private SerializedProperty stringValueProperty;
    private FieldInfo boolValueFieldInfo;
    private SerializedProperty gameObjectProperty;
    private SerializedProperty enumValue;

    private void OnEnable()
    {
        component = target as CustomComponent;
        stringValueProperty = serializedObject.FindProperty("stringValue");
        boolValueFieldInfo = typeof(CustomComponent).GetField(
            "boolValue", BindingFlags.Instance | BindingFlags.NonPublic);
        gameObjectProperty = serializedObject.FindProperty("go");
        enumValue = serializedObject.FindProperty("enumValue");
    }

    public override void OnInspectorGUI()
```

```
    {
        CustomExample();
    }
    private void CustomExample()
    {
        //public 修饰的字段,可以直接访问和修改其值
        component.intValue = EditorGUILayout.IntField(
            "Int Value", component.intValue);
        //private 修饰的字段,通过序列化属性的方式访问和修改其值
        stringValueProperty.stringValue = EditorGUILayout.TextField(
            "String Value", stringValueProperty.stringValue);
        //private 修饰的字段,通过反射的方式访问和修改其值
        boolValueFieldInfo.SetValue(component, EditorGUILayout.Toggle(
            "Bool Value", (bool)boolValueFieldInfo.GetValue(component)));
        EditorGUILayout.PropertyField(gameObjectProperty);
        enumValue.enumValueIndex = EditorGUILayout.Popup(
            "Enum Value", enumValue.enumValueIndex, enumValue.enumNames);
    }
}
```

如图3-1所示,自定义编辑器已经将组件的序列化属性绘制在检视面板中,但是此时修改它们的值并不会被保存,因为当使用自定义编辑器修改组件的序列化属性时,需要调用相应的方法来应用这些修改并保存,而调用这些方法之前需要先检测检视面板或序列化属性是否发生了变更。

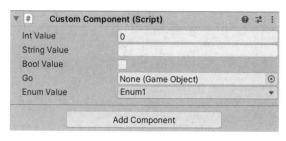

图 3-1 自定义检视面板

3.1.2 如何检测和应用修改

判断序列化属性是否发生变化的方法有多种,可以使用 GUI 类中的 changed 属性,如果任何控件的值发生了变更,则 changed 属性的返回值为 true。也可以使用 EditorGUI 类中的 BeginChangeCheck()方法来开始一块代码块以检测 GUI 的变更,调用 EndChangeCheck()方法结束代码块,如果在该代码块中 GUI 发生了变更,则 EndChangeCheck()方法的返回值将为 true。除此之外,也可以直接对比控件交互修改后的值与属性值是否一致,代码如下:

```
//第 3 章/CustomComponentEditor.cs
```

```csharp
public override void OnInspectorGUI()
{
    ChangeCheckExample();
}
private void ChangeCheckExample()
{
    //开启一块代码块以检测 GUI 是否变更
    EditorGUI.BeginChangeCheck();
    //public 修饰的字段，可以直接访问和修改其值
    component.intValue = EditorGUILayout.IntField(
        "Int Value", component.intValue);
    //如果在代码块中 GUI 发生了变更，则返回值为 true
    if (EditorGUI.EndChangeCheck())
    {
        Debug.Log("IntValue 发生变更");
    }
    //private 修饰的字段，通过序列化属性的方式访问和修改其值
    //接收控件的返回值
    string newStringValue = EditorGUILayout.TextField(
        "String Value", stringValueProperty.stringValue);
    //对比是否一致，如果不一致，则更新
    if (newStringValue != stringValueProperty.stringValue)
    {
        stringValueProperty.stringValue = newStringValue;
        Debug.Log("StringValue 发生变更");
    }
    //private 修饰的字段，通过反射的方式访问和修改其值
    boolValueFieldInfo.SetValue(component, EditorGUILayout.Toggle(
        "Bool Value", (bool)boolValueFieldInfo.GetValue(component)));
    EditorGUILayout.PropertyField(gameObjectProperty);
    enumValue.enumValueIndex = EditorGUILayout.Popup(
        "Enum Value", enumValue.enumValueIndex, enumValue.enumNames);
    //有任何控件发生了变更
    if (GUI.changed)
    {
        Debug.Log("GUI 发生了变更");
    }
}
```

当检测到发生变更后，应该调用序列化对象中的 ApplyModifiedProperties()方法来应用修改，应用后，还需要调用 EditorUtility 中的 SetDirty()方法将对象标记为"脏"，也就是表示它发生了变更，此时再使用快捷键 Ctrl+S 才最终完成保存操作，代码如下：

```csharp
//第 3 章/CustomComponentEditor.cs
public override void OnInspectorGUI()
{
```

```
    ApplyModificationExample();
}
private void ApplyModificationExample()
{
    //public 修饰的字段，可以直接访问和修改其值
    component.intValue = EditorGUILayout.IntField(
        "Int Value", component.intValue);
    //private 修饰的字段，通过序列化属性的方式访问和修改其值
    stringValueProperty.stringValue = EditorGUILayout.TextField(
        "String Value", stringValueProperty.stringValue);
    //private 修饰的字段，通过反射的方式访问和修改其值
    boolValueFieldInfo.SetValue(component, EditorGUILayout.Toggle(
        "Bool Value", (bool)boolValueFieldInfo.GetValue(component)));
    EditorGUILayout.PropertyField(gameObjectProperty);
    enumValue.enumValueIndex = EditorGUILayout.Popup(
        "Enum Value", enumValue.enumValueIndex, enumValue.enumNames);

    //有任何控件发生了变更
    if (GUI.changed)
    {
        //应用修改
        serializedObject.ApplyModifiedProperties();
        //标记为"脏"
        EditorUtility.SetDirty(component);
    }
}
```

3.1.3 编辑器操作的撤销与恢复

如果想要使编辑器中的操作支持撤销和恢复，则需要用到 Undo 类中的方法，常用的几种方法和作用见表 3-2。

表 3-2　Undo 类中的方法和作用

方　　法	作　　用
RecordObject()	记录对象的状态
AddComponent()	为游戏物体添加组件并针对这一操作注册撤销操作
RegisterCreatedObjectUndo()	针对新创建的对象注册撤销操作
DestoryObjectImmediate()	销毁对象并注册撤销操作，以便能够重新创建该对象
SetTransformParent()	更改 Transform 的父级，并注册撤销操作

1. RecordObject()

RecordObject()是最常用的方法，当在编辑器中与控件交互修改某个序列化属性的值时，为了使这次修改操作可撤销，在赋新值之前需要先调用该方法来记录属性的状态，这样在修

改之后使用快捷键 Ctrl+Z 便可撤销本次修改操作，恢复之前的值。

此方法的代码如下，参数 objectToUndo 表示的是要修改的对象的引用，name 表示的是在撤销历史记录中记录的本次操作的名称。

```
public static void RecordObject (Object objectToUndo, string name);
```

以 CustomComponent 组件中的 intValue 为例，使用一个新的 int 值记录控件被编辑后的值，对比新旧值是否一致，如果不一致，则在更新值之前调用 RecordObject()方法，第 1 个参数用于传入对象的引用，第 2 个参数用于传入该项操作的命名，代码如下：

```
//第3章/CustomComponentEditor.cs

public override void OnInspectorGUI()
{
    RecordObjectExample();
}
private void RecordObjectExample()
{
    int newIntValue = EditorGUILayout.IntField(
        "Int Value", component.intValue);
    if (newIntValue != component.intValue)
    {
        Undo.RecordObject(component, "Change Int Value");
        component.intValue = newIntValue;
        serializedObject.ApplyModifiedProperties();
        EditorUtility.SetDirty(component);
    }
}
```

本次操作的名称记录在历史记录中，打开 Edit 菜单可以查看，如图 3-2 所示。

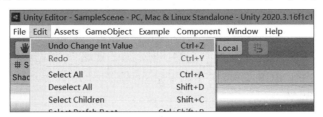

图 3-2　Undo Change Int Value

2. AddComponent()

当在编辑器脚本中为某个游戏物体添加组件时，为了使添加组件这项操作支持撤销与恢复，需要使用 Undo 类中的 AddComponent()方法，而非 GameObject 中的 AddComponent()方法。方法的返回值是添加的新组件，执行撤销操作后，新添加的组件会被销毁。

此方法的代码如下，参数 gameObject 表示要添加组件的游戏物体，type 则是要添加的组件的类型。

```
public static Component AddComponent(GameObject gameObject, Type type);
```

在示例脚本中创建一个 GUI 按钮,当单击该按钮时,使用该方法为组件所在的游戏物体添加一个 BoxCollider 组件。这时如果在编辑器中使用快捷键 **Ctrl+Z** 执行撤销操作,则这个 BoxCollider 组件将会被销毁,代码如下:

```
//第3章/CustomComponentEditor.cs

public override void OnInspectorGUI()
{
    AddComponentExample();
}
private void AddComponentExample()
{
    if (GUILayout.Button("Add BoxCollider"))
    {
        Undo.AddComponent(component.gameObject,
            typeof(BoxCollider));
    }
}
```

操作历史记录如图 3-3 所示。

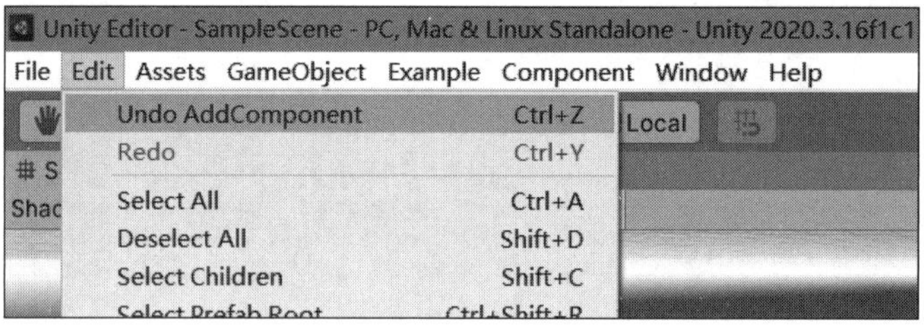

图 3-3　Undo AddComponent

3. RegisterCreatedObjectUndo()

当在编辑器脚本中创建一个新的游戏物体时,为了使这项创建操作支持撤销与恢复,需要在创建游戏物体之后,调用 Undo 类中的 RegisterCreatedObjectUndo()方法,以便为新创建的游戏物体注册撤销操作。

此方法的代码如下,参数 objectToUndo 表示新创建的游戏物体,当执行撤销操作时,该物体节点将被销毁,name 表示在历史记录中记录的本次操作的名称。

```
public static void RegisterCreatedObjectUndo (Object objectToUndo, string name);
```

在示例脚本中创建一个 GUI 按钮,当单击该按钮时,在场景中创建一个新的游戏物体,并调用该方法注册撤销操作。这时如果在编辑器中使用快捷键 **Ctrl+Z** 执行撤销操作,则这个新建的物体节点将被销毁,代码如下:

```
//第 3 章/CustomComponentEditor.cs

public override void OnInspectorGUI()
{
    RegisterCreatedObjectUndoExample();
}
private void RegisterCreatedObjectUndoExample()
{
    if (GUILayout.Button("Create New GameObject"))
    {
        GameObject go = new GameObject();
        Undo.RegisterCreatedObjectUndo(go,
            "Create New GameObject");
    }
}
```

操作历史记录如图 3-4 所示。

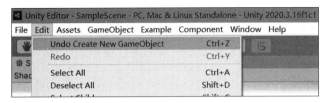

图 3-4　Undo Create New GameObject

4. DestroyObjectImmediate()

当在编辑器脚本中销毁一个对象时，为了使这项销毁操作支持撤销与恢复，需要使用 Undo 类中的 DestoryObjectImmediate()方法，而非 Object 中的 DestroyImmediate()方法。销毁的对象被保存在撤销缓存区中，以便在执行撤销操作时能够重新创建该对象。

此方法的代码如下，参数 objectToUndo 表示要销毁的对象。

```
public static void DestroyObjectImmediate (Object objectToUndo);
```

同样地，在示例脚本中创建一个 GUI 按钮，当单击该按钮时，使用该方法将当前的对象销毁。这时如果在编辑器中使用快捷键 Ctrl+Z 执行撤销操作，则被销毁的对象将被重建，代码如下：

```
//第 3 章/CustomComponentEditor.cs

public override void OnInspectorGUI()
{
    DestroyObjectImmediateExample();
}
private void DestroyObjectImmediateExample()
{
    if (GUILayout.Button("Destroy"))
```

```
    {
        Undo.DestroyObjectImmediate(component);
    }
}
```

操作历史记录如图 3-5 所示。

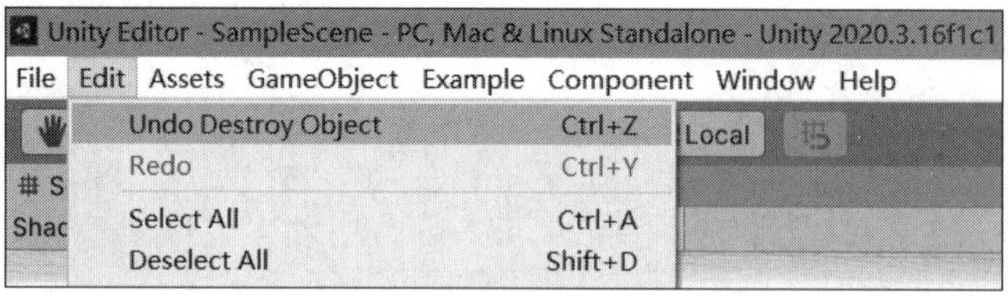

图 3-5 Undo Destroy Object

5. SetTransformParent()

当在编辑器脚本中修改某个 Transform 组件的父级时，为了使这项修改操作支持撤销与恢复，需要使用 Undo 类中的 SetTransformParent() 方法，而非 Transform 中的 SetParent() 方法。

此方法的代码如下，参数 transform 表示要修改父级的 Transform 组件，newParent 表示指定的父级，name 则是在操作历史记录中记录的本次操作的名称。

```
public static void SetTransformParent (Transform transform, Transform newParent, string name);
```

同样地，在示例脚本中创建一个 GUI 按钮，当单击该按钮时，调用该方法将当前游戏物体的父级修改为根级。这时如果在编辑器中使用快捷键 Ctrl+Z 执行撤销操作，则当前游戏物体的父级将由根级恢复为操作之前的父级，代码如下：

```
//第 3 章/CustomComponentEditor.cs

public override void OnInspectorGUI()
{
    SetTransformParentExample();
}
private void SetTransformParentExample()
{
    if (GUILayout.Button("Set As Root"))
    {
        Undo.SetTransformParent(component.transform,
            null, "Change Transform Parent");
    }
}
```

操作历史记录如图 3-6 所示。

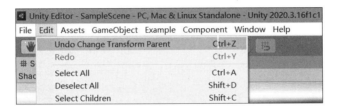

图 3-6 Undo Change Transform Parent

3.1.4 实现 DoTween 动画参数的编辑

DoTween 是一个免费开源的高效动画工具，相信大部分开发者或多或少会用到它。使用 DoTween 工具可以实现 Transform 组件的移动、旋转和缩放动画，在调用相关方法时，可以设置动画时长、延时、起始值、目标值等参数。

本节针对 Transform 封装一个动画组件，并在它的编辑器类中自定义这些动画参数的绘制，为它们提供控件，以便进行交互修改。

首先定义移动、旋转和缩放动画相关的参数，以旋转动画为例，toggle 表示动画的开关，startValue 与 endValue 分别表示起始值和目标值，duration 表示动画的时长，delay 表示动画的延时时长，ease 则表示 DoTween 动画的类型，不同的类型对应不同的动画效果。以上这些参数在移动动画和缩放动画中都包含，而 mode 是旋转动画中特有的字段，表示旋转的类型，代码如下：

```
//第3章/RotateAnimation.cs

using System;
using UnityEngine;
using DG.Tweening;

[Serializable]
public class RotateAnimation
{
    public bool toggle;
    public Vector3 startValue;
    public Vector3 endValue;
    public float duration = 1f;
    public float delay = 0f;
    public Ease ease = Ease.Linear;
    public RotateMode mode = RotateMode.Fast;
}
```

然后创建 Transform 的 Tween 动画组件，move、rotate、scale 分别表示移动动画、旋转动画、缩放动画，代码如下：

```
//第3章/TransformTweenAnimation.cs
```

```
using UnityEngine;

public class TransformTweenAnimation : MonoBehaviour
{
    public MoveAnimation move;
    public RotateAnimation rotate;
    public ScaleAnimation scale;
}
```

将组件挂载于游戏物体，默认检视面板如图 3-7 所示。

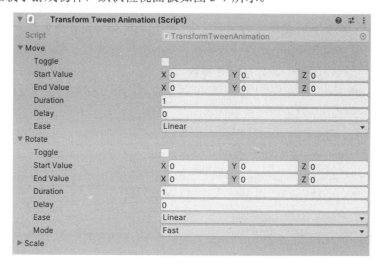

图 3-7　TransformTweenAnimation 组件的默认检视面板

为 TransformTweenAnimation 组件创建自定义编辑器类，注意需要为编辑器类添加 CustomEditor 特性，否则它无法生效。在 OnGUI()方法中添加菜单栏，使用按钮控制各动画的开关，开启水平布局，将这些按钮水平排列，还需要使用 Undo 类中的 RecordObject()方法记录属性的状态，以便用户使用快捷键 Ctrl+Z 与 Ctrl+Y 实现编辑操作的撤销与恢复，代码如下：

```
//第 3 章/TransformTweenAnimationEditor.cs

using UnityEngine;
using UnityEditor;

[CustomEditor(typeof(TransformTweenAnimation))]
public class TransformTweenAnimationEditor : Editor
{
    private TransformTweenAnimation animation;
    private readonly static float alpha = .4f;

    private void OnEnable()
```

```csharp
{
    animation = target as TransformTweenAnimation;
}
public override void OnInspectorGUI()
{
    OnMenuGUI();
}
private void OnMenuGUI()
{
    GUILayout.BeginHorizontal();
    Color cacheColor = GUI.color;
    Color alphaColor = new Color(cacheColor.r,
        cacheColor.g, cacheColor.b, alpha);
    //Move
    GUI.color = animation.move.toggle ? cacheColor : alphaColor;
    if (GUILayout.Button(EditorGUIUtility.IconContent("MoveTool"),
        "ButtonLeft", GUILayout.Width(25f)))
    {
        Undo.RecordObject(target, "Move Toggle");
        animation.move.toggle = !animation.move.toggle;
        EditorUtility.SetDirty(target);
    }
    //Rotate
    GUI.color = animation.rotate.toggle ? cacheColor : alphaColor;
    if (GUILayout.Button(EditorGUIUtility.IconContent("RotateTool"),
        "ButtonMid", GUILayout.Width(25f)))
    {
        Undo.RecordObject(target, "Rotate Toggle");
        animation.rotate.toggle = !animation.rotate.toggle;
        EditorUtility.SetDirty(target);
    }
    //Scale
    GUI.color = animation.scale.toggle ? cacheColor : alphaColor;
    if (GUILayout.Button(EditorGUIUtility.IconContent("ScaleTool"),
        "ButtonRight", GUILayout.Width(25f)))
    {
        Undo.RecordObject(target, "Scale Toggle");
        animation.scale.toggle = !animation.scale.toggle;
        EditorUtility.SetDirty(target);
    }
    GUI.color = cacheColor;
    GUILayout.EndHorizontal();
}
}
```

接下来为各动画的参数添加交互修改的控件，实现动画参数的编辑，仍然以旋转动画为例，代码如下：

```csharp
//第3章/TransformTweenAnimationEditor.cs

readonly static float labelWidth = 60f;
readonly static GUIContent duration = new GUIContent("Duration", "动画时长");
readonly static GUIContent delay = new GUIContent("Delay", "延时时长");
readonly static GUIContent from = new GUIContent("From", "初始值");
readonly static GUIContent to = new GUIContent("To", "目标值");
readonly static GUIContent ease = new GUIContent("Ease");
readonly static GUIContent rotateMode = new GUIContent("Mode", "旋转模式");

public override void OnInspectorGUI()
{
    OnMenuGUI();
    OnMoveAnimationGUI();
    GUILayout.Space(3f);
    OnRotateAnimationGUI();
    GUILayout.Space(3f);
    OnScaleAnimationGUI();
}
private void OnRotateAnimationGUI()
{
    if (animation.rotate.toggle)
    {
        GUILayout.BeginHorizontal("Badge");
        {
            GUILayout.BeginVertical();
            GUILayout.Space(40f);
            GUILayout.Label(EditorGUIUtility.IconContent("RotateTool"));
            GUILayout.EndVertical();

            GUILayout.BeginVertical();
            {
                //Duration、Delay
                GUILayout.BeginHorizontal();
                GUILayout.Label(duration, GUILayout.Width(labelWidth));
                var newDuration = EditorGUILayout.FloatField(
                    animation.rotate.duration);
                if (newDuration != animation.rotate.duration)
                {
                    Undo.RecordObject(target, "Rotate Duration");
                    animation.rotate.duration = newDuration;
                    EditorUtility.SetDirty(target);
                }

                GUILayout.Label(delay, GUILayout.Width(labelWidth - 20f));
                var newDelay = EditorGUILayout.FloatField(
```

```
    animation.rotate.delay);
if (newDelay != animation.rotate.delay)
{
    Undo.RecordObject(target, "Rotate Delay");
    animation.rotate.delay = newDelay;
    EditorUtility.SetDirty(target);
}
GUILayout.EndHorizontal();

//From
GUILayout.BeginHorizontal();
GUILayout.Label(from, GUILayout.Width(labelWidth));
Vector3 newStartValue = EditorGUILayout.Vector3Field(
    GUIContent.none, animation.rotate.startValue);
if (newStartValue != animation.rotate.startValue)
{
    Undo.RecordObject(target, "Rotate From");
    animation.rotate.startValue = newStartValue;
    EditorUtility.SetDirty(target);
}
GUILayout.EndHorizontal();

//To
GUILayout.BeginHorizontal();
GUILayout.Label(to, GUILayout.Width(labelWidth));
Vector3 newEndValue = EditorGUILayout.Vector3Field(
    GUIContent.none, animation.rotate.endValue);
if (newEndValue != animation.rotate.endValue)
{
    Undo.RecordObject(target, "Rotate To");
    animation.rotate.endValue = newEndValue;
    EditorUtility.SetDirty(target);
}
GUILayout.EndHorizontal();

//Rotate Mode
GUILayout.BeginHorizontal();
GUILayout.Label(rotateMode, GUILayout.Width(labelWidth));
var newRotateMode = (RotateMode)EditorGUILayout.EnumPopup(
    animation.rotate.mode);
if (newRotateMode != animation.rotate.mode)
{
    Undo.RecordObject(target, "Rotate Mode");
    animation.rotate.mode = newRotateMode;
    EditorUtility.SetDirty(target);
}
GUILayout.EndHorizontal();
```

```
            //Ease
            GUILayout.BeginHorizontal();
            GUILayout.Label(ease, GUILayout.Width(labelWidth));
            var newEase = (Ease)EditorGUILayout.EnumPopup(
                animation.rotate.ease);
            if (newEase != animation.rotate.ease)
            {
                Undo.RecordObject(target, "Rotate Ease");
                animation.rotate.ease = newEase;
                EditorUtility.SetDirty(target);
            }
            GUILayout.EndHorizontal();
        }
        GUILayout.EndVertical();
    }
    GUILayout.EndHorizontal();
}
```

最终效果如图 3-8 所示。

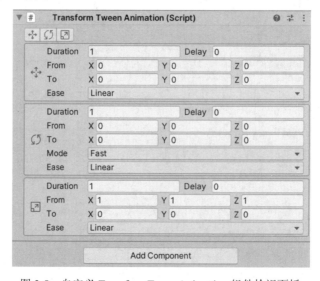

图 3-8　自定义 TransformTweenAnimation 组件检视面板

3.1.5　如何自定义预览窗口

预览窗口是指当选中模型、动画或音频等类型的资源时，在检视窗口的下方出现的小窗口，如图 3-9 所示。在默认情况下，在组件对象的检视窗口中是没有预览窗口的，如果想开启预览窗口，则需要重写 Editor 类中的虚方法 HasPreviewGUI()，当方法的返回值为 true 时就可以打开预览窗口，而预览窗口中的内容需要重写 OnPreviewGUI() 方法实现。

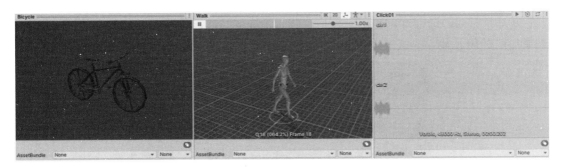

图 3-9 预览窗口

除此之外，Editor 类中还有两个重要的与预览窗口相关的虚方法，即 GetPreviewTitle() 和 OnPreviewSettings()，前者可以设置预览窗口的标题，后者可以在标题栏的右侧添加设置相关的控件。

仍然以 CustomComponent 组件为例，让 HasPreviewGUI()方法在程序运行时返回值 true，并在 OnPreviewGUI()方法中将各序列化属性的值绘制出来，代码如下：

```
//第 3 章/CustomComponentEditor.cs

public override bool HasPreviewGUI()
{
    return EditorApplication.isPlaying;
}
public override GUIContent GetPreviewTitle()
{
    return new GUIContent("这里是窗口的标题");
}
public override void OnPreviewSettings()
{
    GUILayout.Button("Button1", EditorStyles.toolbarButton);
    GUILayout.Button("Button2", EditorStyles.toolbarButton);
}
public override void OnPreviewGUI(Rect r, GUIStyle background)
{
    GUILayout.Label(string.Format("Int Value: {0}",
        example.intValue));
    GUILayout.Label(string.Format("String Value: {0}",
        stringValueProperty.stringValue));
    GUILayout.Label(string.Format("Bool Value: {0}",
        (bool)boolValueFieldInfo.GetValue(example)));
    GUILayout.Label(string.Format("Go Value: {0}",
        gameObjectProperty.objectReferenceValue));
    GUILayout.Label(string.Format("Enum Value: {0}",
        enumValue.enumNames[enumValue.enumValueIndex]));
}
```

程序运行后，选中 CustomComponent 组件所在的物体，查看 Inspector 中的预览窗口，结果如图 3-10 所示。

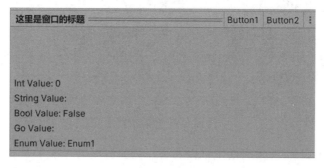

图 3-10　自定义预览窗口

OnPreviewGUI()方法的第 1 个参数为 Rect 类型，它表示预览窗口的矩形区域。通常情况下，在根据矩形区域绘制内容时，需要使用 GUI 或 EditorGUI 中的方法，GUILayout 类和 EditorGUILayout 类中的方法不再适用。GUI 类位于 Unity Engine 命名空间中，EditorGUI 位于 Unity Editor 命名空间中，与 GUILayout 和 EditorGUILayout 一样，它们有不同的适用范围。

1. EditorGUI

EditorGUI 类中包含 3 个公开的静态变量，其中 actionKey 表示当前是否按下了 Ctrl 键，这对于使用快捷键实现某些功能是十分有用的，它在 macOS 系统中对应的是 Command 键。indentLevel 用于控制字段标签的缩进级别。showMixedValue 表示编辑多个不同值的外观，例如，在层级窗口中选中多个游戏物体，如果这些游戏物体的 Transform 组件拥有不同的坐标、旋转和缩放值，则在检视面板中这些变量的值将使用"—"来表示，如图 3-11 所示。

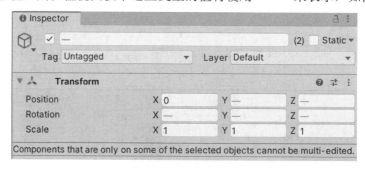

图 3-11　批量编辑

EditorGUI 类中包含的静态方法见表 3-3。

表 3-3　EditorGUI 类中的静态方法

方　　法	作　　用
BeginChangeCheck()	开启一个新代码块来检查 GUI 是否发生了变更，与 EndChangeCheck 配合使用
BeginDisabledGroup()	创建一组禁用的控件，与 EndDisabledGroup 配合使用，表示组内的控件无法进行交互
BeginFoldoutHeaderGroup()	创建一个折叠栏，与 EndFoldoutHeaderGroup 配合使用
BeginProperty()	创建一个属性封装器，可用于使常规 GUI 控件与 SerializedProperty 配合使用
BoundsField()	创建一个用于编辑 Bounds 类型字段的控件
BoundsIntField()	创建一个用于编辑 BoundsInt 类型字段的控件
CanCacheInspectorGUI()	确定能否缓存 SerializedProperty 的检视面板 GUI
ColorField()	创建一个用于编辑 Color 类型字段的控件
CurveField()	创建一个用于编辑 AnimationCurve 类型字段的控件
DelayedDoubleField()	创建一个用于编辑 double 类型字段的延迟类型输入框
DelayedFloatField()	创建一个用于编辑 float 类型字段的延迟类型输入框
DelayedIntField()	创建一个用于编辑 int 类型字段的延迟类型输入框
DelayedTextField()	创建一个用于编辑 string 类型字段的延迟类型输入框
DoubleField()	创建一个用于编辑 double 类型字段的输入框
DrawPreviewTexture()	在矩形内绘制纹理
DrawRect()	在当前编辑器窗口中的指定位置以指定大小绘制一个着色的矩形
DrawTextureAlpha()	在矩形内绘制纹理的 Alpha 通道
DropdownButton()	创建一个用于打开下拉列表样式的按钮
DropShadowLabel()	绘制带有投影的文本
EndChangeCheck()	结束由 BeginChangeCheck 开启的代码块，并检查 GUI 是否发生了变更
EndDisabledGroup()	结束由 BeginDisabledGroup 开始的禁用组
EndFoldoutHeaderGroup()	结束由 BeginFoldoutHeaderGroup 开启的折叠栏
EndProperty()	结束由 BeginProperty 开始的属性封装器
EnumFlagsField()	创建一个用于编辑位掩码的控件
EnumPopup()	创建一个用于编辑枚举类型字段的控件
FloatField()	创建一个用于编辑 float 类型字段的输入框
FocusTextInControl()	将键盘焦点移动到指定的文本字段，并开始编辑内容
Foldout()	创建一个折叠栏
GetPropertyHeight()	获取 PropertyField 控件所需的高度
GradientField()	创建一个用于编辑 Gradient 类型字段的控件
HandlePrefixLabel()	创建一个显示在特定控件前的文本

续表

方法	作用
HelpBox()	创建一个带有发送给用户的消息的帮助框
InspectorTitlebar()	创建一个类似于 Inspector 窗口的标题栏
IntField()	创建一个用于编辑 int 类型字段的输入框
IntPopup()	创建一个用于弹出下拉列表的控件
IntSlider()	创建一个用于编辑 int 类型字段的滑动条
LabelField()	创建一个文本（用于显示只读信息）
LayerField()	创建一个用于编辑 Layer 层级的控件
LongField()	创建一个用于编辑 long 类型字段的输入框
MaskField()	创建一个用于编辑掩码的控件
MinMaxSlider()	创建一个滑动条
MultiFloatField()	创建一个用于编辑多个 float 类型字段的控件，用于在同一行中编辑多个浮点值
MultiIntField()	创建一个用于编辑多个 int 类型字段的控件，用于在同一行中编辑多个整数
MultiPropertyField()	创建一个用于编辑多个序列化属性的控件
ObjectField()	创建一个用于编辑对象的控件（通过拖曳对象或使用对象选择器分配对象）
PasswordField()	创建一个用于编辑密码类型文本的输入框
Popup()	创建一个用于弹出下拉列表的控件
PrefixLabel()	创建一个显示在特定控件前的文本
ProgressBar()	创建一个进度条
PropertyField()	创建一个用于编辑可序列化属性的控件
RectField()	创建一个用于编辑 Rect 类型字段的控件
RectIntField()	创建一个用于编辑 RectInt 类型字段的控件
SelectableLabel()	创建一个可选择的文本（用于显示可复制粘贴的只读信息）
Slider()	创建一个滑动条
TagField()	创建一个用于编辑物体 Tag 标签的控件
TextArea()	创建一个用于编辑 string 类型字段的文本域
TextField()	创建一个用于编辑 string 类型字段的输入框
Toggle()	创建一个开关
ToggleLeft()	创建一个开关，开关位于左侧，标签紧随其右
Vector2Field()	创建一个用于编辑 Vector2 类型字段的控件
Vector2IntField()	创建一个用于编辑 Vector2Int 类型字段的控件
Vector3Field()	创建一个用于编辑 Vector3 类型字段的控件
Vector3IntField()	创建一个用于编辑 Vector3Int 类型字段的控件
Vector4Field()	创建一个用于编辑 Vector4 类型字段的控件

2. 预览 Transform 组件的 DoTween 动画

PreviewRenderUtility 是用于预览渲染的功用类，它可以在预览窗口中提供一个预览场景，可以在预览场景中添加游戏物体，还可以通过调整场景中的灯光、相机等参数调整预览效果，常用的方法见表 3-4。

表 3-4 PreviewRenderUtility 类的常用方法

方　　法	作　　用	方　　法	作　　用
BeginPreview()	开始预览	EndStaticPreview()	结束静态预览
BeginStaticPreview()	开始静态预览	AddSingleGO()	在预览场景中添加游戏物体
EndPreview()	结束预览	Cleanup()	释放相关资源

在 3.1.4 节中创建了一个 TransformTweenAnimation 组件，实现了 Transform 组件的 DoTween 动画参数的编辑，本节使用 EditorGUI 类与 PreviewRenderUtility 类在该组件的预览窗口中实现动画的预览。

首先在 OnEnable() 方法中对当前动画组件所挂载的游戏物体进行实例化，将实例通过预览渲染功用类中的 AddSingleGO() 方法添加到预览场景中。在 OnDisable() 方法中，需要调用预览渲染功用类中的 Cleanup() 方法来释放相关资源，否则在其析构函数执行时将会报错，代码如下：

```
//第3章/TransformTweenAnimationEditor.cs

private TransformTweenAnimation animation;
private PreviewRenderUtility previewRenderUtility;
private GameObject previewInstance;
private bool isPlaying;

private void OnEnable()
{
    animation = target as TransformTweenAnimation;

    previewInstance = Instantiate(animation.gameObject,
        Vector3.zero, Quaternion.identity);
    previewRenderUtility = new PreviewRenderUtility(true);
    previewRenderUtility.AddSingleGO(previewInstance);
    EditorApplication.update += Update;
}
private void OnDisable()
{
    EditorApplication.update -= Update;

    previewRenderUtility.Cleanup();
    previewRenderUtility = null;
    DestroyImmediate(previewInstance);
```

```
        previewInstance = null;
    }
    private void Update()
    {
        if (isPlaying)
        {
            Repaint();
        }
    }
```

然后重写 Editor 的 OnPreviewSettings()方法,在其中提供动画开始预览和停止预览的按钮控件,代码如下:

```
//第3章/TransformTweenAnimationEditor.cs
public override void OnPreviewSettings()
{
    if (GUILayout.Button(EditorGUIUtility.IconContent(
        !isPlaying ? "PlayButton" : "PauseButton",
        !isPlaying ? "|开始预览" : "|停止预览"),
        EditorStyles.toolbarButton))
    {
        isPlaying = !isPlaying;
        //单击"播放"按钮,开始预览动画
        if (isPlaying)
        {
            Sequence sequence = DOTween.Sequence();
            //移动动画
            if (animation.move.toggle)
            {
                sequence.Insert(0f, previewInstance.transform
                    .DOMove(animation.move.endValue,
                        animation.move.duration)
                    .SetEase(animation.move.ease)
                    .SetDelay(animation.move.delay)
                    .From(animation.move.startValue));
            }
            //旋转动画
            if (animation.rotate.toggle)
            {
                sequence.Insert(0f, previewInstance.transform
                    .DORotate(animation.rotate.endValue,
                        animation.rotate.duration, animation.rotate.mode)
                    .SetEase(animation.rotate.ease)
                    .SetDelay(animation.rotate.delay)
                    .From(animation.rotate.startValue));
            }
```

```csharp
                //缩放动画
                if (animation.scale.toggle)
                {
                    sequence.Insert(0f, previewInstance.transform
                        .DOScale(animation.scale.endValue,
                            animation.scale.duration)
                        .SetEase(animation.scale.ease)
                        .SetDelay(animation.scale.delay)
                        .From(animation.scale.startValue));
                }
                //循环播放
                sequence.SetLoops(-1);
                //开始预览
                DG.DOTweenEditor.DOTweenEditorPreview
                    .PrepareTweenForPreview(sequence);
                DG.DOTweenEditor.DOTweenEditorPreview.Start();
            }
            //单击"停止"按钮，停止预览动画
            else
            {
                DG.DOTweenEditor.DOTweenEditorPreview.Stop();
                //重置坐标、旋转、缩放
                previewInstance.transform.position = Vector3.zero;
                previewInstance.transform.rotation = Quaternion.identity;
                previewInstance.transform.localScale = Vector3.one;
            }
        }
    }
}
```

当在预览窗口中拖曳鼠标时，可以旋转预览场景中的相机视角，滚动鼠标滚轮可以调整相机与预览实例的距离。在预览窗口中获取输入事件，代码如下：

```csharp
//第3章/TransformTweenAnimationEditor.cs

private Vector2 dragRot;
private float distance = 5f;

//在预览窗口中的输入事件
private void OnPreviewGUIInput(Rect r, ref Vector2 dragRot,
    ref float distance)
{
    int hashCode = GetType().GetHashCode();
    int controlID = GUIUtility.GetControlID(hashCode, FocusType.Passive);
    switch (Event.current.GetTypeForControl(controlID))
    {
        case EventType.MouseDown:
            //鼠标按下并且是预览窗口区域
```

```
            if (r.Contains(Event.current.mousePosition))
            {
                GUIUtility.hotControl = controlID;
                Event.current.Use();
                //鼠标移出屏幕外后从另一侧移入
                EditorGUIUtility.SetWantsMouseJumping(1);
            }
            break;
        case EventType.MouseUp:
            if (GUIUtility.hotControl == controlID)
                GUIUtility.hotControl = 0;
            EditorGUIUtility.SetWantsMouseJumping(0);
            break;
        case EventType.MouseDrag:
            if (GUIUtility.hotControl == controlID)
            {
                dragRot -= Event.current.delta /
                    Mathf.Min(r.width, r.height) * 140f;
                Event.current.Use();
                GUI.changed = true;
            }
            break;
        case EventType.ScrollWheel:
            distance += Event.current.delta.y * .1f;
            distance = Mathf.Clamp(distance, 3f, 30f);
            Event.current.Use();
            GUI.changed = true;
            break;
    }
}
```

最后重写OnPreviewGUI()方法，定义预览窗口中预览的内容，代码如下：

```
//第3章/TransformTweenAnimationEditor.cs

public override void OnPreviewGUI(Rect r, GUIStyle background)
{
    //获取各类型输入
    OnPreviewGUIInput(r, ref dragRot, ref distance);
    //重绘事件
    if (Event.current.type == EventType.Repaint)
    {
        //开启预览区域
        previewRenderUtility.BeginPreview(r, background);
        //调整相机旋转、坐标
        Camera camera = previewRenderUtility.camera;
        camera.transform.rotation = Quaternion.Euler(
```

```
            new Vector3(-dragRot.y, -dragRot.x, 0));
        camera.transform.position = camera.transform.forward * -distance;
        //相机的相关设置
        EditorUtility.SetCameraAnimateMaterials(camera, true);
        camera.cameraType = CameraType.Preview;
        camera.enabled = false;
        camera.clearFlags = CameraClearFlags.Skybox;
        camera.fieldOfView = 30f;
        camera.farClipPlane = 50f;
        camera.nearClipPlane = 2f;
        //相机渲染
        camera.Render();
        //预览
        previewRenderUtility.EndAndDrawPreview(r);

        EditorGUI.LabelField(new Rect(r.x, r.y, r.width, 20f),
            string.Format("Position: {0}",
                previewInstance.transform.position));
        EditorGUI.LabelField(new Rect(r.x, r.y + 20f, r.width, 20f),
            string.Format("Rotation: {0}",
                previewInstance.transform.eulerAngles));
        EditorGUI.LabelField(new Rect(r.x, r.y + 40f, r.width, 20f),
            string.Format("Scale: {0}",
                previewInstance.transform.localScale));
    }
}
```

结果如图 3-12 所示。

3.1.6 扩展默认组件的检视面板

本章中介绍了使用 Editor 类和 CustomEditor 特性为组件创建编辑器类，重写虚方法 OnInspectorGUI()可以自定义组件的检视面板，通过保留 base.OnInspectorGUI()的调用可以在保留原有的检视面板内容的基础上扩展检视面板。这种方式用于开发者自己创建的组件没有任何问题，但是如果用于 Unity 默认已有的组件，则可能会出现丢失原有检视面板内容的情况。以 RectTransform 组件为例，该组件原有的检视面板如图 3-13 所示。

为 RectTransform 组件创建一个自定义编辑器类，代码如下：

```
//第 3 章/RectTransformEditorExample.cs

using UnityEngine;
using UnityEditor;

[CustomEditor(typeof(RectTransform))]
public class RectTransformEditorExample : Editor
{
```

```csharp
public override void OnInspectorGUI()
{
    base.OnInspectorGUI();
}
```

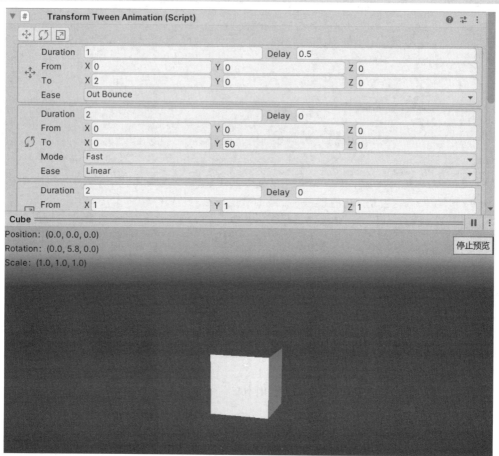

图 3-12 预览 Transform 组件的 DoTween 动画

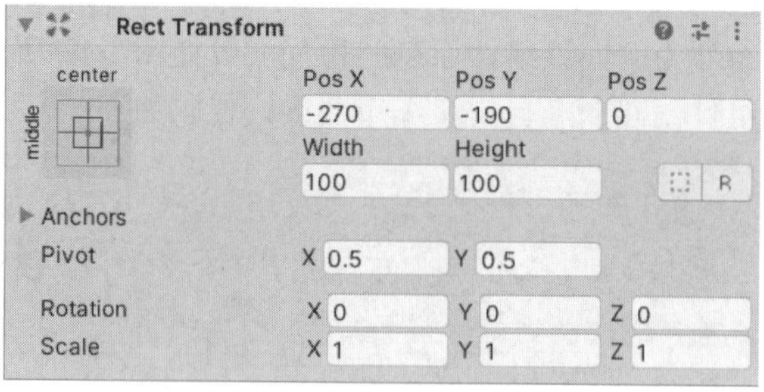

图 3-13 RectTransform 组件原有的检视面板

结果如图 3-14 所示，可以看到，尽管保留了 base.OnInspectorGUI()的调用，但是原有的检视面板内容还是发生了变化。

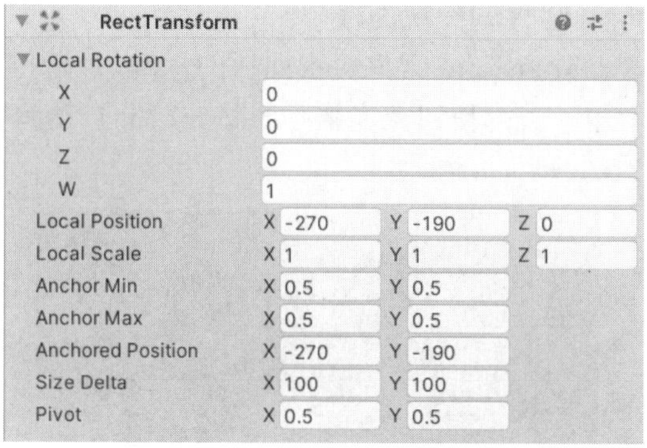

图 3-14　RectTransform 组件创建自定义编辑器类后的检视面板

造成这种现象的原因在于，Unity 本身已经实现了 RectTransform 组件的编辑器类，类名为 RectTransformEditor，只不过该类不是公开可见的，通过反编译工具可以看到该类的实现，如图 3-15 所示。

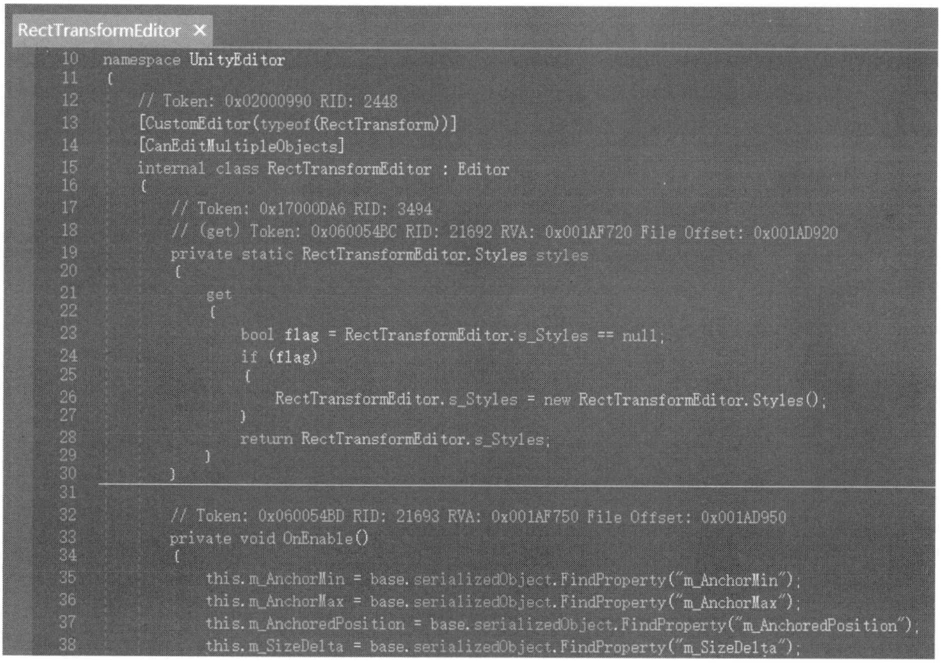
RectTransformEditor ×

```
namespace UnityEditor
{
    // Token: 0x02000990 RID: 2448
    [CustomEditor(typeof(RectTransform))]
    [CanEditMultipleObjects]
    internal class RectTransformEditor : Editor
    {
        // Token: 0x17000DA6 RID: 3494
        // (get) Token: 0x060054BC RID: 21692 RVA: 0x001AF720 File Offset: 0x001AD920
        private static RectTransformEditor.Styles styles
        {
            get
            {
                bool flag = RectTransformEditor.s_Styles == null;
                if (flag)
                {
                    RectTransformEditor.s_Styles = new RectTransformEditor.Styles();
                }
                return RectTransformEditor.s_Styles;
            }
        }

        // Token: 0x060054BD RID: 21693 RVA: 0x001AF750 File Offset: 0x001AD950
        private void OnEnable()
        {
            this.m_AnchorMin = base.serializedObject.FindProperty("m_AnchorMin");
            this.m_AnchorMax = base.serializedObject.FindProperty("m_AnchorMax");
            this.m_AnchoredPosition = base.serializedObject.FindProperty("m_AnchoredPosition");
            this.m_SizeDelta = base.serializedObject.FindProperty("m_SizeDelta");
```

图 3-15　RectTransformEditor

因此，开发者在创建 RectTransform 组件的编辑器类后，Unity 默认的该组件的编辑器类便失去了作用。那么该如何解决该问题？可以通过 Editor 类中的 CreateEditor()方法创建指定类型的编辑器实例，在 OnInspectorGUI()方法中调用编辑器实例的 OnInspectorGUI()方法即可。

1. 扩展 RectTransform 组件的检视面板

本节介绍一个扩展 RectTransform 组件检视面板的示例，该示例在 RectTransform 组件原本的检视面板的基础上增加了一个按钮，当执行该按钮时会自动为 RectTransform 组件设置锚点，锚点的样式如图 3-16 所示。

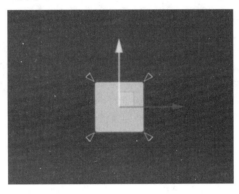

图 3-16　设置锚点

编辑器类在 OnEnable()方法中做初始化操作，获取 RectTransformEditor 类型，通过类型创建编辑器实例，相应地，在 OnDisable()方法中销毁编辑器实例。由于在 RectTransformEditor 类中有 OnSceneGUI()方法的实现，而该方法是私有的，因此可以通过反射的方式调用该方法，代码如下：

```csharp
//第3章/RectTransformEditorExample.cs

using System.Linq;
using System.Reflection;

using UnityEngine;
using UnityEditor;

[CustomEditor(typeof(RectTransform))]
public class RectTransformEditorExample : Editor
{
    private RectTransform rt;
    private Editor instance;
    private MethodInfo onSceneGUI;
    private static readonly object[] emptyArray = new object[0];

    private void OnEnable()
    {
        rt = target as RectTransform;
        //获取 RectTransformEditor 类型
```

```csharp
        var editorType = Assembly.GetAssembly(typeof(Editor)).GetTypes()
            .FirstOrDefault(m => m.Name == "RectTransformEditor");
        //创建指定类型的编辑器实例
        instance = CreateEditor(targets, editorType);
        //由于 OnSceneGUI 方法是私有的,因此可以通过反射方式调用
        onSceneGUI = editorType.GetMethod("OnSceneGUI",
            BindingFlags.Instance | BindingFlags.Static
            | BindingFlags.Public | BindingFlags.NonPublic);
    }
    private void OnSceneGUI()
    {
        if (instance)
            onSceneGUI.Invoke(instance, emptyArray);
    }
    private void OnDisable()
    {
        if (instance != null)
            DestroyImmediate(instance);
    }
    public override void OnInspectorGUI()
    {
        instance.OnInspectorGUI();
        OnAnchorSetHelperGUI();
    }

    //锚点设置工具
    private void OnAnchorSetHelperGUI()
    {
        EditorGUILayout.Space();
        Color color = GUI.color;
        GUI.color = Color.cyan;
        GUILayout.Label("锚点工具", EditorStyles.boldLabel);
        GUI.color = color;
        if (GUILayout.Button("Auto Anchor"))
        {
            Undo.RecordObject(rt, "Auto Anchor");
            RectTransform prt = rt.parent as RectTransform;
            Vector2 anchorMin = new Vector2(
                rt.anchorMin.x + rt.offsetMin.x / prt.rect.width,
                rt.anchorMin.y + rt.offsetMin.y / prt.rect.height);
            Vector2 anchorMax = new Vector2(
                rt.anchorMax.x + rt.offsetMax.x / prt.rect.width,
                rt.anchorMax.y + rt.offsetMax.y / prt.rect.height);
            rt.anchorMin = anchorMin;
            rt.anchorMax = anchorMax;
            rt.offsetMin = rt.offsetMax = Vector2.zero;
        }
```

```
        }
    }
```

RectTransform 组件扩展后的检视面板如图 3-17 所示。

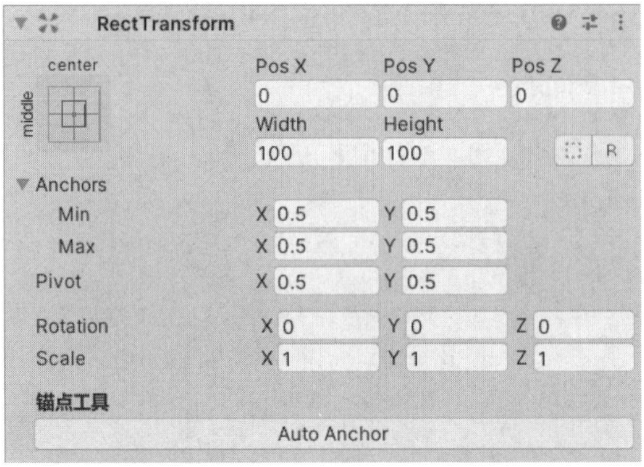

图 3-17　扩展后 RectTransform 组件的检视面板

2. 扩展 Transform 组件的检视面板

本节介绍一个扩展 Transform 组件检视面板的示例，在 Transform 组件原本的检视面板的基础上增加一个按钮，当单击该按钮时，将该组件所在游戏物体的层级路径复制到系统粘贴板中。

获取游戏物体层级路径可以通过依次获取物体父级，将它们存储到一个列表中，然后倒序遍历列表进行字符串拼接实现。

通过反编译工具可以发现，Unity 中默认的 Transform 组件的编辑器类名称为 TransformInspector，如图 3-18 所示。

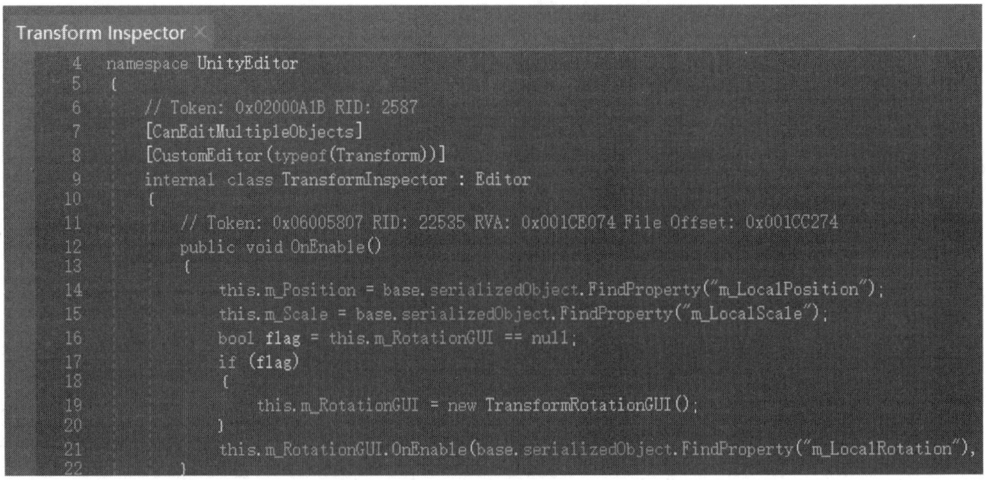
Transform Inspector

```
namespace UnityEditor
{
    // Token: 0x02000A1B RID: 2587
    [CanEditMultipleObjects]
    [CustomEditor(typeof(Transform))]
    internal class TransformInspector : Editor
    {
        // Token: 0x06005807 RID: 22535 RVA: 0x001CE074 File Offset: 0x001CC274
        public void OnEnable()
        {
            this.m_Position = base.serializedObject.FindProperty("m_LocalPosition");
            this.m_Scale = base.serializedObject.FindProperty("m_LocalScale");
            bool flag = this.m_RotationGUI == null;
            if (flag)
            {
                this.m_RotationGUI = new TransformRotationGUI();
            }
            this.m_RotationGUI.OnEnable(base.serializedObject.FindProperty("m_LocalRotation"),
        }
```

图 3-18　TransformInspector

因此在创建Transform组件的编辑器实例时，需要通过类名TransformInspector获取类型，然后根据该类型创建编辑器实例，代码如下：

```csharp
//第3章/TransformEditor.cs

using System;
using System.Text;
using System.Linq;
using System.Reflection;
using System.Collections.Generic;

using UnityEngine;
using UnityEditor;

[CustomEditor(typeof(Transform))]
public class TransformEditor : Editor
{
    private Editor instance;

    private void OnEnable()
    {
        //获取TransformInspector类型
        Type editorType = Assembly.GetAssembly(typeof(Editor)).GetTypes()
            .FirstOrDefault(m => m.Name == "TransformInspector");
        //创建编辑器实例
        instance = CreateEditor(targets, editorType);
    }
    private void OnDisable()
    {
        //销毁编辑器实例
        if (instance != null)
            DestroyImmediate(instance);
    }

    public override void OnInspectorGUI()
    {
        instance.OnInspectorGUI();
        GUILayout.Space(10f);
        if (GUILayout.Button("Copy Full Path"))
        {
            List<Transform> tfs = new List<Transform>();
            Transform tf = target as Transform;
            tfs.Add(tf);
            while (tf.parent)
            {
                tf = tf.parent;
                tfs.Add(tf);
```

```
            }
            StringBuilder sb = new StringBuilder();
            sb.Append(tfs[tfs.Count - 1].name);
            for (int i = tfs.Count - 2; i >= 0; i--)
            {
                sb.Append("/" + tfs[i].name);
            }
            GUIUtility.systemCopyBuffer = sb.ToString();
        }
    }
}
```

Transform 组件扩展后的检视面板如图 3-19 所示。

如图 3-20 所示,在 Hierarchy 窗口中选中 Sphere 游戏物体,在 Inspector 窗口中单击 Copy Full Path 按钮后,系统粘贴板中的内容为 GameObject/Cube/Sphere。

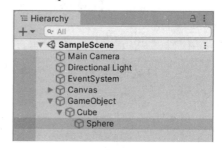

图 3-19　扩展 Transform 组件的检视面板　　　图 3-20　在 Hierarchy 窗口选中 Sphere 游戏物体

3.2　PropertyDrawer

CustomEditor 特性允许开发者自定义一个组件的检视面板如何绘制,而 CustomPropertyDrawer 特性可以自定义特定类型的属性如何绘制,使其更加直观和易于使用,它可以配合 PropertyAttribute 进行使用。在了解 CustomPropertyDrawer 特性之前,先了解一些 Unity 内置的 PropertyDrawer。

3.2.1　内置的 PropertyDrawer

1. Range

RangeAttribute 用于脚本中 float 或 int 类型变量,使用后,float 或 int 类型的变量将被序列化到检视面板中,并且不再是默认的那样通过输入框控件交互修改,而是通过滑动条进行修改,这样可以限制变量的取值范围。构造函数的代码如下,参数 min 表示允许的最小值,max 表示允许的最大值。

```
public RangeAttribute(float min, float max);
```

示例代码如下：

```
using UnityEngine;

public class PropertyDrawerExample : MonoBehaviour
{
    [Range(4f, 10f)]
    [SerializeField] private float radius = 5f;
}
```

结果如图 3-21 所示。

图 3-21　RangeAttribute

2. Min

MinAttribute 同样用于 float 或 int 类型变量，其作用是限制变量的最小值，被序列化到检视面板中后默认通过输入框控件交互修改，但是输入的值如果小于限定的最小值，则会被修改为最小值。构造函数的代码如下，参数 min 表示允许的最小值。

```
public MinAttribute(float min);
```

示例代码如下：

```
using UnityEngine;

public class PropertyDrawerExample : MonoBehaviour
{
    [Min(4f)]
    [SerializeField] private float radius = 5f;
}
```

结果如图 3-22 所示。

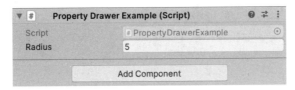

图 3-22　MinAttribute

3. Multiline

MultilineAttribute 用于字符串类型变量，其作用是通过指定行数的输入框控件实现字符串内容的交互修改。构造函数的代码如下，参数 lines 表示行数，默认值为 3。

```
public MultilineAttribute(int lines);
```

示例代码如下：

```
using UnityEngine;

public class PropertyDrawerExample : MonoBehaviour
{
    [Multiline(5), SerializeField]
    private string text = "Hello World.\r\nToday is a good day.";
}
```

结果如图 3-23 所示。

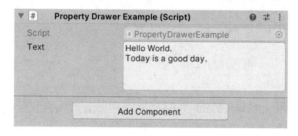

图 3-23　MultilineAttribute

4. TextArea

TextAreaAttribute 同样用于字符串类型变量，使字符串具有高度可调且可滚动的文本区域。构造函数的代码如下，参数 minLines 表示最小行数，maxLines 表示最大行数，它们的默认值均为 3。

```
public TextAreaAttribute(int minLines, int maxLines);
```

示例代码如下：

```
using UnityEngine;

public class PropertyDrawerExample : MonoBehaviour
{
    [TextArea, SerializeField]
    private string text = "Hello World.\r\nToday is a good day.";
}
```

结果如图 3-24 所示。

5. ColorUsage

ColorUsageAttribute 用于 Color 类型变量，其作用是提供可指定是否为 HDR 类型的颜色选择器。构造函数的代码如下，参数 showAlpha 表示是否显示 alpha 通道，hdr 表示是否为 HDR 颜色类型，默认值为 false。

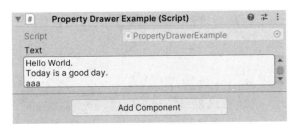

图 3-24　TextAreaAttribute

```
public ColorUsageAttribute(bool showAlpha, bool hdr);
```

示例代码如下：

```
using UnityEngine;

public class PropertyDrawerExample : MonoBehaviour
{
    [ColorUsage(true, true), SerializeField]
    private Color color = Color.white;
}
```

结果如图 3-25 所示。

图 3-25　ColorUsageAttribute

6. GradientUsage

GradientUsageAttribute 用于 Gradient 类型变量，可以指定渐变的颜色是否为 HDR 类型及颜色空间。构造函数的代码如下，参数 hdr 表示是否为 HDR 颜色类型，colorSpace 表示渐变使用的颜色空间。

```
public GradientUsageAttribute(bool hdr, ColorSpace colorSpace);
```

示例代码如下：

```
using UnityEngine;

public class PropertyDrawerExample : MonoBehaviour
{
    [GradientUsage(false, ColorSpace.Gamma)]
    [SerializeField] private Gradient gradient;
}
```

结果如图 3-26 所示。

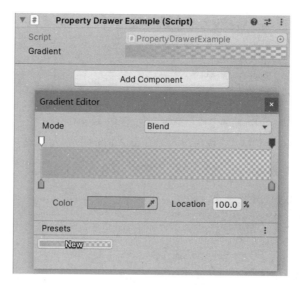

图 3-26　GradientUsageAttribute

3.2.2　内置的 DecoratorDrawer

DecoratorDrawer 类似于 PropertyDrawer，但它不绘制属性，而是纯粹基于从其对应的 PropertyAttribute 中获取的数据来绘制装饰元素。虽然 DecoratorDrawer 从概念上讲并不意味着要与特定字段相关联，但其属性仍需放在组件中的字段上。与 PropertyDrawer 属性不同的是，同一字段上可以有多个 DecoratorDrawer 属性。如果 DecoratorDrawer 属性放在列表或数组的字段上，则该装饰器只会在此数组前显示一次，而不是针对每个数组元素显示。

1. Space

SpaceAttribute 用于在检视面板中增加一些间隙，也就是插入一个空白元素。构造函数的代码如下，参数 height 表示插入的间隙的高度，默认为 8 像素。

```
public SpaceAttribute(float height);
```

示例代码如下：

```
using UnityEngine;

public class PropertyDrawerExample : MonoBehaviour
{
    [SerializeField] private float radius = 5f;
    [Space(10f)]
    [SerializeField] private string text = "Hello World.";
    [SerializeField] private int intValue = 2;
}
```

结果如图 3-27 所示。

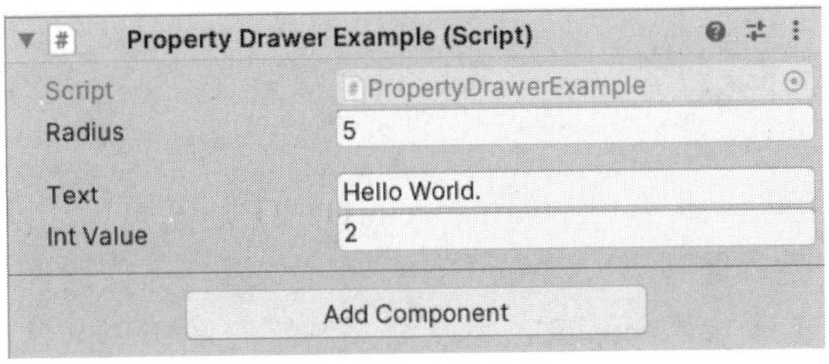

图 3-27　SpaceAttribute

2. Header

HeaderAttribute 用于在某些字段上添加标题。构造函数的代码如下，参数 header 表示标题的内容。

```
public HeaderAttribute(string header);
```

示例代码如下：

```
using UnityEngine;

public class PropertyDrawerExample : MonoBehaviour
{
    [Header("***相关变量")]
    [SerializeField] private float radius = 5f;
    [SerializeField] private string text = "Hello World.";
}
```

结果如图 3-28 所示。

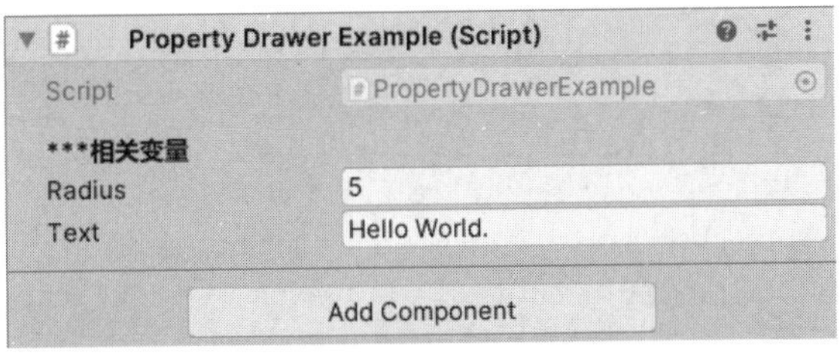

图 3-28　HeaderAttribute

3.2.3　如何创建自定义 PropertyDrawer

以一个 TimeAttribute 为例，使用该特性可以将一个 float 类型的以秒为单位的描述时间的字段在检视面板上以 00:00:00 的时间格式进行显示。首先创建这个类并继承 PropertyAttribute，它用于 float 类型的字段，使用 AttributeUsage 定义使用的范围，然后创建对应的绘制类，继承 PropertyDrawer，并为其添加 CustomPropertyDrawer 特性，代码如下：

```
[AttributeUsage(AttributeTargets.Field)]
public sealed class TimeAttribute : PropertyAttribute { }
```

```
[CustomPropertyDrawer(typeof(TimeAttribute))]
public sealed class TimePropertyDrawer : PropertyDrawer { }
```

如同在 Editor 中重写虚方法 OnInspectorGUI()实现自定义绘制检视面板一样，CustomPropertyDrawer 需要重写虚方法 OnGUI()来定义属性如何绘制。

OnGUI()方法的第 1 个参数 position 表示绘制该字段的矩形区域，此参数为 Rect 类型，因此在添加控件时使用 EditorGUI 中的方法。第 2 个参数 property 表示对应的序列化属性，第 3 个参数 label 表示描述该字段的文本内容。

在方法中判断如果属性的类型不是 float 类型，则使用 EditorGUI 类中的 HelpBox()方法提示只能用于 float 类型字段。在将秒数转换为 00:00:00 时间格式的字符串时，首先将 float 转换为 int 类型，然后将 int 类型值对 3600 取整就是最终的小时数，对 3600 取余后再对 60 取整就是最终的分钟数，对 3600 取余后再对 60 取余就是最终的秒数，代码如下：

```
//第 3 章/TimePropertyDrawer.cs

using System;
using UnityEditor;
using UnityEngine;

[CustomPropertyDrawer(typeof(TimeAttribute))]
public sealed class TimePropertyDrawer : PropertyDrawer
{
    public override void OnGUI(Rect position,
        SerializedProperty property, GUIContent label)
    {
        if (property.propertyType == SerializedPropertyType.Float)
        {
            property.floatValue = EditorGUI.FloatField(
                new Rect(position.x, position.y,
                    position.width * .6f, position.height),
                label, property.floatValue);
            EditorGUI.LabelField(new Rect(
                position.x + position.width * .6f, position.y,
                position.width * .4f, position.height),
                GetTimeFormat(property.floatValue));
        }
        else
        {
            //如果将 Time 用于 float 之外的类型，则会报错提示
            EditorGUI.HelpBox(position,
                "只支持 float 类型字段", MessageType.Error);
        }
    }
    //转换为时间格式字符串
```

```csharp
    private string GetTimeFormat(float time)
    {
        //取整获得总共的秒数
        int l = Convert.ToInt32(time);
        //小时数是秒数对 3600 取整
        int hours = l / 3600;
        //分钟数是秒数对 3600 取余后再对 60 取整
        int minutes = l % 3600 / 60;
        //最终的秒数是对 3600 取余后再对 60 取余
        int seconds = l % 3600 % 60;
        return string.Format("({0:D2}:{1:D2}:{2:D2})",
            hours, minutes, seconds);
    }
}
```

使用该特性的示例代码如下：

```csharp
//第 3 章/PropertyDrawerExample.cs
using UnityEngine;
public class PropertyDrawerExample : MonoBehaviour
{
    [Time, SerializeField]
    private float duration = 596f;
    [Time, SerializeField]
    private int delay = 2;
}
```

结果如图 3-29 所示。

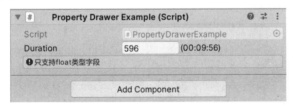

图 3-29　TimeAttribute

PropertyDrawer 类中除了 OnGUI()方法外，还有一个重要的虚方法 GetPropertyHeight()，该方法用于指定绘制字段的 GUI 像素高度，在 2020.3.16f1c1 的 Unity 版本中，它的默认高度是 18 像素。如果需要绘制的内容过多，则可以换行进行绘制，这时需要重写该方法，设置更大的高度。

1. ColorPropertyDrawer

Color 类型的属性在检视面板中默认通过 ColorField 控件进行交互修改，以 Image 组件中的 Color 属性为例，如图 3-30 所示。

本节通过 PropertyDrawer 自定义 Color 属性的绘制,在 ColorField 后面增加两个控件,内容分别是表示颜色的十六进制字符串和颜色的 Alpha 值,如图 3-31 所示。这两个值原本需要单击 ColorFiled 区域打开编辑 Color 的浮窗后进行修改,如图 3-32 所示,添加到这里后可以直接在检视面板中进行编辑。

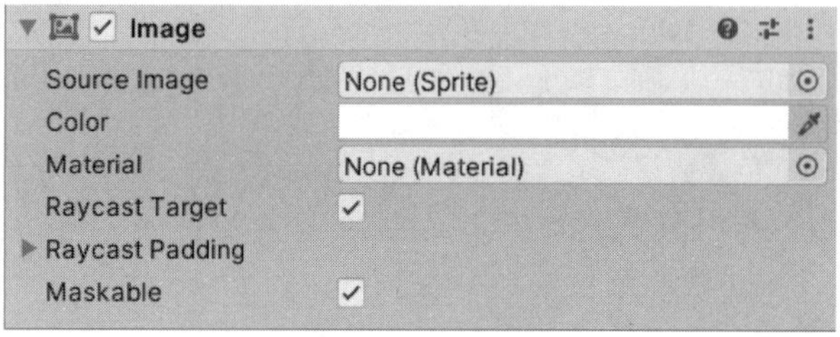

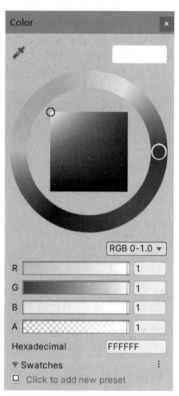

图 3-30 Image 组件

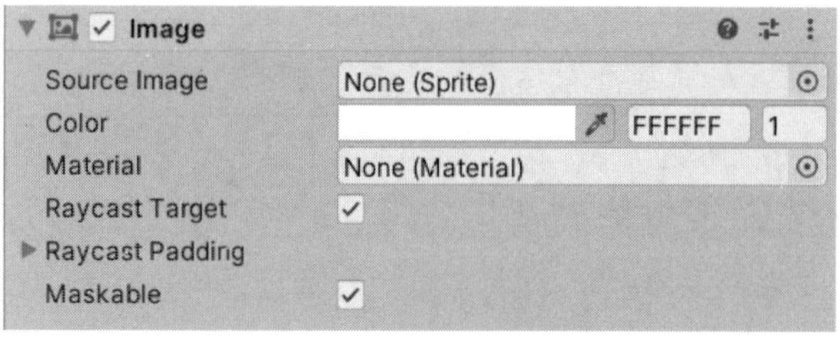

图 3-31 ColorPropertyDrawer

图 3-32 Color 编辑浮窗

代码如下:

```
//第3章/ColorPropertyDrawer.cs

using System;
using System.Globalization;

using UnityEngine;
using UnityEditor;

[CustomPropertyDrawer(typeof(Color))]
public class ColorPropertyDrawer : PropertyDrawer
{
    private const float spacing = 5f;
    private const float hexWidth = 60f;
```

```
private const float alphaWidth = 32f;

public override void OnGUI(Rect position,
    SerializedProperty property, GUIContent label)
{
    label = EditorGUI.BeginProperty(
        position, label, property);
    position = EditorGUI.PrefixLabel(position,
        GUIUtility.GetControlID(FocusType.Passive), label);

    var indent = EditorGUI.indentLevel;
    EditorGUI.indentLevel = 0;
    float colorWidth = position.width - hexWidth
        - spacing - alphaWidth - spacing;

    Color newColor = EditorGUI.ColorField(new Rect(
        position.x, position.y, colorWidth, position.height),
            property.colorValue);
    if (!newColor.Equals(property.colorValue))
        property.colorValue = newColor;

    //十六进制颜色值字符串
    string hex = EditorGUI.TextField(new Rect(position.x + colorWidth
        + spacing, position.y, hexWidth, position.height),
            ColorUtility.ToHtmlStringRGB(property.colorValue));
    try
    {
        newColor = FromHex(hex, property.colorValue.a);
        if (!newColor.Equals(property.colorValue))
            property.colorValue = newColor;
    }
    catch(Exception e)
    {
        Debug.LogError(e);
    }

    //颜色 Alpha 值
    float newAlpha = EditorGUI.Slider(new Rect(position.x + colorWidth
        + hexWidth + (spacing * 2f), position.y, alphaWidth,
        position.height), property.colorValue.a, 0f, 1f);
    if (!newAlpha.Equals(property.colorValue.a))
        property.colorValue = new Color(property.colorValue.r,
            property.colorValue.g, property.colorValue.b, newAlpha);

    EditorGUI.indentLevel = indent;
    EditorGUI.EndProperty();
}
```

```csharp
//十六进制颜色值字符串转Color值
private static Color FromHex(string hexValue, float alpha)
{
    if (string.IsNullOrEmpty(hexValue))
        return Color.clear;
    if (hexValue[0] == '#')
        hexValue = hexValue.TrimStart('#');
    if (hexValue.Length > 6)
        hexValue = hexValue.Remove(6, hexValue.Length - 6);
    int value = int.Parse(hexValue, NumberStyles.HexNumber);
    int r = value >> 16 & 255;
    int g = value >> 8 & 255;
    int b = value & 255;
    return new Color(r / 255f, g / 255f, b / 255f, alpha);
}
```

2. SpritePropertyDrawer

在示例组件 SpriteExample 中声明一个 Sprite 类型的变量，将其挂载于场景中的游戏物体，为变量赋值一张 Sprite 类型的图片，检视面板如图 3-33 所示，代码如下：

```csharp
using UnityEngine;

public class SpriteExample : MonoBehaviour
{
    public Sprite axeSprite;
}
```

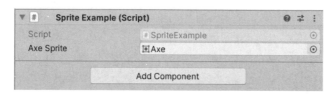

图 3-33　Sprite 属性默认的检视面板

Sprite 类型的属性在检视面板中默认以纹理图标和图片名称的形式显示在 ObjectField 类型的控件中。如果想要查看图片的具体内容，则需要单击控件区域，在 Project 窗口中定位该图片以便进行查看。

本节通过 PropertyDrawer 自定义 Sprite 类型属性的绘制，其目的是可以在检视面板中直观地看到图片的具体内容，如图 3-34 所示。

代码如下：

```
//第 3 章/SpritePropertyDrawer.cs
```

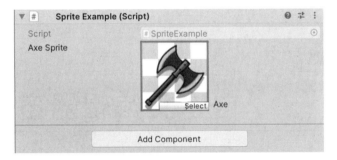

图 3-34　自定义 Sprite 类型属性绘制

```
using UnityEngine;
using UnityEditor;

[CustomPropertyDrawer(typeof(Sprite))]
public class SpritePropertyDrawer : PropertyDrawer
{
    public override float GetPropertyHeight(
        SerializedProperty property, GUIContent label)
    {
        return 110f;
    }

    public override void OnGUI(Rect position,
        SerializedProperty property, GUIContent label)
    {
        label = EditorGUI.BeginProperty(
            position, label, property);
        position = EditorGUI.PrefixLabel(position,
            GUIUtility.GetControlID(FocusType.Passive), label);

        EditorGUI.ObjectField(new Rect(position.x, position.y, 100f, 100f),
            property, typeof(Sprite), GUIContent.none);

        if (property.objectReferenceValue != null)
        {
            Rect spriteNameRect = new Rect(position.x + 105f,
                position.y + 35f,position.width - 105f, position.height);
            EditorGUI.LabelField(spriteNameRect,
                property.objectReferenceValue.name);
        }
        EditorGUI.EndProperty();
    }
}
```

第 4 章 自定义编辑器窗口

本章主要介绍如何自定义编辑器窗口，涉及 EditorWindow、PopupWindow、GenericMenu 和 ScriptableWizard 等相关类。除此之外，本章还将介绍如何扩展 Unity 默认的编辑器窗口，例如 Project 窗口、Hierarchy 窗口等。通过本章的学习，读者将能够为自己的项目开发并提供更多的工具。

4.1 如何创建新的编辑器窗口

Unity 的使用是在不同的编辑器窗口中进行的，例如 Scene、Game、Project、Hierarchy、Inspector、Console 等，除了这些默认的编辑器窗口，开发者可以创建新的编辑器窗口并自定义其中的内容，创建的新的编辑器窗口类需要继承 EditorWindow。

4.1.1 打开新创建的编辑器窗口

打开编辑器窗口需要一个菜单路径，例如打开控制台窗口的菜单路径为 Window/General/Console，所以首先要为新创建的编辑器窗口使用 MenuItem 提供一个窗口入口。调用 EditorWindow 类中的静态方法 GetWindow() 可以获取指定类型的窗口实例，该方法具有多个重载，代码如下：

```
public static T GetWindow<T>() where T : EditorWindow;
public static T GetWindow<T>(bool utility) where T : EditorWindow;
public static T GetWindow<T>(bool utility, string title) where T : EditorWindow;
public static T GetWindow<T>(string title) where T : EditorWindow;
public static T GetWindow<T>(string title, bool focus) where T : EditorWindow;
public static T GetWindow<T>(bool utility, string title, bool focus) where T : EditorWindow;
public static T GetWindow<T>(params Type[] desiredDockNextTo) where T : EditorWindow;
public static T GetWindow<T>(string title, params Type[] desiredDockNextTo) where T : EditorWindow;
```

```
public static T GetWindow<T>(string title, bool focus, params Type[] desiredDockNextTo) where T : EditorWindow;
```

参数 utility 表示窗口的类型，当设为 true 时可创建浮动的窗口，当设为 false 时可以创建正常窗口，title 表示窗口的标题，focus 表示是否聚焦窗口，desiredDockNextTo 表示窗口试图停靠到其上的 EditorWindow 类型集合。

窗口的标题默认以编辑器窗口类的类型名称命名，也可以通过 titleContent 进行修改。窗口的最小尺寸默认为 100×100，最大尺寸默认为 4000×4000，可以分别通过 minSize、maxSize 进行修改。最终调用窗口类中的 Show() 方法即可打开窗口，代码如下：

```
//第 4 章/MyEditorWindow.cs

using UnityEngine;
using UnityEditor;

public class MyEditorWindow : EditorWindow
{
    [MenuItem("Example/My Editor Window")]
    public static void Open()
    {
        MyEditorWindow window = GetWindow<MyEditorWindow>();
        window.titleContent = new GUIContent("窗口标题");
        window.minSize = new Vector2(300f, 300f);
        window.maxSize = new Vector2(1920f, 1080f);
        window.Show();
    }
}
```

4.1.2 定义编辑器窗口中的 GUI 内容

EditorWindow 类中的回调方法及作用见表 4-1，其中 OnGUI() 是最常用的回调方法，自定义编辑器窗口中的 GUI 内容如何绘制全部在 OnGUI() 方法中进行定义。

表 4-1 EditorWindow 类中的回调方法及作用

回 调 方 法	作　　用
OnEnable()	当窗口打开时被调用
OnGUI()	在此处实现自定义 GUI 内容
OnFocus()	当聚焦窗口时被调用
OnLostFucus()	与 OnFocus() 对应，当窗口失去焦点时被调用
OnHierarchyChange()	当 Hierarchy 层级窗口中结构发生变化时被调用
OnInspectorUpdate()	以每秒 10 帧的速度被调用，为 Inspector 检视窗口提供机会进行更新
OnProjectChange()	当 Project 窗口发生变化时被调用

续表

回调方法	作用
OnSelectionChange()	当选择发生变更时被调用
OnValidate()	当该脚本被加载时被调用
OnDisable()	当窗口关闭时被调用
OnDestroy()	当窗口关闭时被调用,晚于 OnDisable()执行

示例代码如下:

```csharp
//第4章/MyEditorWindow.cs

using UnityEngine;
using UnityEditor;

public class MyEditorWindow : EditorWindow
{
    [MenuItem("Example/My Editor Window")]
    public static void Open()
    {
        MyEditorWindow window = GetWindow<MyEditorWindow>();
        window.titleContent = new GUIContent("窗口标题");
        window.minSize = new Vector2(300f, 300f);
        window.maxSize = new Vector2(1920f, 1080f);
        window.Show();
    }
    private void OnEnable()
    {
        Debug.Log("OnEnable");
    }
    private void OnGUI()
    {
        GUILayout.Button("Button");
    }
    private void OnFocus()
    {
        Debug.Log("OnFocus");
    }
    private void OnLostFocus()
    {
        Debug.Log("OnLostFocus");
    }
    private void OnHierarchyChange()
    {
        Debug.Log("OnHierarchyChang");
    }
    private void OnInspectorUpdate()
```

```
    {
        Debug.Log("OnInspecotrUpdate");
    }
    private void OnProjectChange()
    {
        Debug.Log("OnProjectChang");
    }
    private void OnSelectionChange()
    {
        Debug.Log("OnSelectionChang");
    }
    private void OnValidate()
    {
        Debug.Log("OnValidate");
    }
    private void OnDisable()
    {
        Debug.Log("OnDisable");
    }
    private void OnDestroy()
    {
        Debug.Log("OnDestroy");
    }
}
```

在 OnGUI()方法中绘制了一个按钮控件，如图 4-1 所示。

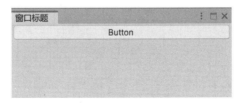

图 4-1　My Editor Window

4.1.3　如何创建弹出窗口

1. PopupWindow

PopupWindow 弹出窗口不可以被拖动，也无法调节大小，并且还会在失去焦点时自动关闭。打开弹出窗口需要调用 PopupWindow 中的静态方法 Show()，代码如下：

```
public static void Show (Rect activatorRect, PopupWindowContent windowContent);
```

参数 activatorRect 表示打开弹出窗口的按钮的矩形区域，windowContent 表示弹出窗口内容的实例，抽象基类为 PopupWindowContent，在它的派生类中需要实现抽象方法 OnGUI()，与 EditorWindow 一样，可以在该方法中定义窗口绘制的内容。除此之外，在该类中还有其他 3 个虚方法，见表 4-2。

表 4-2　PopupWindowContent 类中的虚方法

方　　法	详　　解
GetWindowSize()	返回值为 Vector2 类型，它定义了弹出窗口的尺寸大小
OnOpen()	打开事件，当弹出窗口打开时被调用
OnClose()	关闭事件，当弹出窗口关闭时被调用

示例代码如下：

```
//第 4 章/MyEditorWindow.cs

using UnityEngine;
using UnityEditor;

public class MyEditorWindow : EditorWindow
{
    [MenuItem("Example/My Editor Window")]
    public static void Open()
    {
        MyEditorWindow window = GetWindow<MyEditorWindow>();
        window.titleContent = new GUIContent("窗口标题");
        window.minSize = new Vector2(300f, 300f);
        window.maxSize = new Vector2(1920f, 1080f);
        window.Show();
    }
    private Rect examplePupupWindowRect;
    private void OnGUI()
    {
        if (GUILayout.Button("PopupWindow Example"))
            PopupWindow.Show(examplePupupWindowRect,
                new ExamplePopupWindowContent(
                    new Vector2(position.width - 6f, 100f)));
        if (Event.current.type == EventType.Repaint)
            examplePupupWindowRect = GUILayoutUtility.GetLastRect();
    }
}

public class ExamplePopupWindowContent : PopupWindowContent
{
    //窗口尺寸
    private Vector2 windowSize;
    //滚动值
    private Vector2 scroll;
```

```csharp
public ExamplePopupWindowContent(Vector2 windowSize)
{
    this.windowSize = windowSize;
}
public override Vector2 GetWindowSize()
{
    return windowSize;
}
public override void OnOpen()
{
    Debug.Log("打开示例弹出窗口");
}
public override void OnClose()
{
    Debug.Log("关闭示例弹出窗口");
}
public override void OnGUI(Rect rect)
{
    scroll = EditorGUILayout.BeginScrollView(scroll);
    GUILayout.BeginHorizontal();
    GUILayout.Toggle(false, "Toggle1");
    GUILayout.Toggle(true, "Toggle2");
    GUILayout.Toggle(true, "Toggle3");
    GUILayout.EndHorizontal();
    GUILayout.Button("Button1");
    GUILayout.Button("Button2");
    GUILayout.Button("Button3");
    EditorGUILayout.EndScrollView();
}
}
```

如图 4-2 所示，当单击 PopupWindowExample 按钮时，按钮下方会出现弹出窗口，在弹出窗口中绘制的内容为 ExamplePopupWindowContent 类 OnGUI()方法中定义的内容。

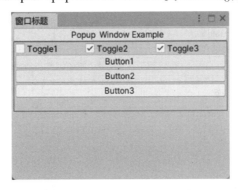

图 4-2　PopupWindowExample

2. GenericMenu

使用 GenericMenu 同样可以创建一个弹出窗口，与 PopupWindow 的区别在于，后者可以在弹出窗口内开启水平或垂直布局，创建各类型的交互控件，还可以灵活地绘制编辑器元素，而前者弹出的窗口只是一个下拉菜单，它包含的公共方法及作用见表 4-3。

表 4-3 Generic Menu 中的公共方法及作用

方法	作用	方法	作用
AddDisableItem()	向菜单中添加一个不可交互项	DropDown()	在指定的矩形区域中显示菜单
AddItem()	向菜单中添加一个交互项	GetItemCount()	获取菜单项的数量
AddSeparator()	向菜单中添加一个分隔符	ShowAsContext()	在鼠标位置显示菜单

AddItem()方法用于添加菜单项，代码如下，参数 content 表示菜单项的路径；on 表示菜单项是否激活，当为 true 时，在菜单项内容前会有勾选标记；func 表示当选中该菜单项时调用的方法。

```
public void AddItem (GUIContent content, bool on, GenericMenu.MenuFunction func);
```

AddSeparator()方法用于添加分隔符，代码如下，如果向一级菜单中添加分隔符，则参数 path 传空字符串，如果向子菜单中添加分隔符，则参数 path 传子菜单的路径。

```
public void AddSeparator (string path);
```

示例代码如下：

```
if (GUILayout.Button("Button"))
{
    GenericMenu gm = new GenericMenu();
    //添加菜单项
    gm.AddItem(new GUIContent("Memu1"), true,
        () =>Debug.Log("Select Menu1"));
    //添加分隔符，参数传空字符串表示在一级菜单中添加分隔符
    gm.AddSeparator(string.Empty);
    //添加不可交互菜单项
    gm.AddDisabledItem(new GUIContent("Memu2"));
    //通过'/'可添加子菜单项
    gm.AddItem(new GUIContent("Menu3/SubMenu1"), false,
        () =>Debug.Log("Select SubMenu1"));
    //在子菜单中添加分隔符
    gm.AddSeparator("Menu3/");
    gm.AddItem(new GUIContent("Menu3/SubMenu2"), false,
        () =>Debug.Log("Select SubMenu2"));
    //显示菜单
    gm.ShowAsContext();
}
```

结果如图 4-3 所示。

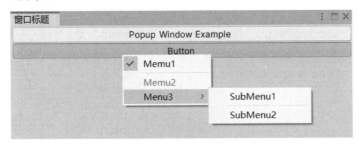

图 4-3　Generic Menu

4.1.4　开发备忘录

本节介绍一个开发备忘录工具示例，该示例对于 EditorWindow、PopupWindow、GenericMenu 中的知识点均有应用，如图 4-4 所示，开发者可以在里面记录开发日志或者待办任务。顶部的 GUI 中有两个控件，左侧是一个下拉按钮，单击时使用 GenericMenu 列举了几种列表排序方法，右侧是一个输入框，它用于支持内容检索。下面的 GUI 中分为左右两部分，左侧是备忘录的列表，右侧是当前选中项的数据详情。

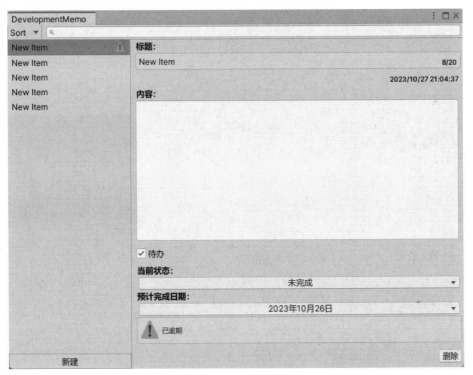

图 4-4　开发备忘录

实现该工具首先需要定义备忘录的数据结构，每项数据包含标题、创建时间、内容、是否为待办、待办是否完成及预计完成时间等字段。备忘录的数据需要被保存在本地缓存中，因此需要数据结构类支持序列化，代码如下：

```csharp
//第 4 章/DevelopmentMemo.cs

[Serializable]
public class DevelopmentMemoItem
{
    //<summary>
    //标题
    //</summary>
    public string title;
    //<summary>
    //创建时间
    //</summary>
    public DateTime createdTime;
    //<summary>
    //内容
    //</summary>
    public string content;
    //<summary>
    //是否为待办
    //</summary>
    public bool todo;
    //<summary>
    //预计完成时间
    //</summary>
    public DateTime estimateCompleteTime;
    //<summary>
    //是否已经完成
    //</summary>
    public bool isCompleted;
    //<summary>
    //是否逾期
    //</summary>
    public bool isOverdue;

    //<summary>
    //计算是否已经逾期
    //</summary>
    public void OverdueCal()
    {
        isOverdue = !isCompleted && (DateTime.Now
            - estimateCompleteTime).Days > 0;
    }
```

```
}
[Serializable]
public class DevelopmentMemoData
{
    public List<DevelopmentMemoItem> list
        = new List<DevelopmentMemoItem>(0);
}
```

备忘录中的每项既可以是一个开发日志,也可以是一个待办任务,待办任务中有任务的预计完成时间,日期的选择通过单击下拉按钮实现,打开弹出窗口,在弹出窗口中选择该任务的预计完成时间。

弹出窗口的大小需要根据预计完成时间的下拉列表按钮的大小进行设定,因此在构造函数中传入 Vector2 类型的参数值用于设定弹出窗口的大小。实现抽象方法 OnGUI(),在里面绘制年、月、日的选择控件,由于弹出窗口的宽度是不定的,所以每行需要绘制的控件数量无法确定,因此通过窗口宽度对每个控件所需的宽度取整来确定每行绘制的控件数量,代码如下:

```
//第4章/DevelopmentMemo.cs

public class DatePopupWindowContent : PopupWindowContent
{
    //弹出窗口的尺寸
    private Vector2 windowSize;
    //滚动值
    private Vector2 scroll;
    private readonly DevelopmentMemoItem item;

    public DatePopupWindowContent(Vector2 windowSize,
        DevelopmentMemoItem item)
    {
        this.windowSize = windowSize;
        this.item = item;
    }

    public override Vector2 GetWindowSize()
    {
        return windowSize;
    }

    public override void OnGUI(Rect rect)
    {
        Color cacheColor = GUI.color;
        scroll = EditorGUILayout.BeginScrollView(scroll);
        GUILayout.Label("年", EditorStyles.boldLabel);
        GUILayout.BeginHorizontal();
```

```csharp
            int currentYear = DateTime.Now.Year;
            for (int i = 0; i < 3; i++)
            {
                GUI.color = currentYear + i == item.estimateCompleteTime.Year
                    ? Color.gray : cacheColor;
                if (GUILayout.Button((currentYear + i).ToString(),
                    GUILayout.Width(60f)))
                {
                    item.estimateCompleteTime
                        = new DateTime(currentYear + i, 1, 1);
                }
                GUI.color = cacheColor;
            }
            GUILayout.EndHorizontal();

            EditorGUILayout.Space();
            GUILayout.Label("月", EditorStyles.boldLabel);
            int monthCountPerRow = Mathf.RoundToInt(rect.width / 53f);
            for (int i = 0; i < 12; i += monthCountPerRow)
            {
                GUILayout.BeginHorizontal();
                for (int j = 0; j < monthCountPerRow; j++)
                {
                    int index = i + j;
                    if (index < 12)
                    {
                        GUI.color = (index + 1)
                            == item.estimateCompleteTime.Month
                            ? Color.gray : cacheColor;
                        if (GUILayout.Button((index + 1).ToString(),
                            GUILayout.Width(50f)))
                        {
                            item.estimateCompleteTime = new DateTime(
                                item.estimateCompleteTime.Year, index + 1, 1);
                            item.OverdueCal();
                        }
                        GUI.color = cacheColor;
                    }
                }
                GUILayout.EndHorizontal();
            }

            EditorGUILayout.Space();
            GUILayout.Label("日", EditorStyles.boldLabel);
            int daysCount = DateTime.DaysInMonth(item.estimateCompleteTime.Year,
                item.estimateCompleteTime.Month);
            int dayCountPerRow = Mathf.RoundToInt(rect.width / 43f);
```

```
            for (int i = 0; i < daysCount; i += dayCountPerRow)
            {
                GUILayout.BeginHorizontal();
                for (int j = 0; j < dayCountPerRow; j++)
                {
                    int index = i + j;
                    if (index < daysCount)
                    {
                        GUI.color = (index + 1)
                            == item.estimateCompleteTime.Day
                            ? Color.gray : cacheColor;
                        if (GUILayout.Button((index + 1).ToString(),
                            GUILayout.Width(40f)))
                        {
                            item.estimateCompleteTime = new DateTime(
                                item.estimateCompleteTime.Year,
                                item.estimateCompleteTime.Month, index + 1);
                            item.OverdueCal();
                        }
                        GUI.color = cacheColor;
                    }
                }
                GUILayout.EndHorizontal();
            }
            GUILayout.EndScrollView();
        }
    }
```

结果如图4-5所示。

当备忘录编辑器窗口打开时，根据缓存文件的路径读取缓存数据，然后通过二进制反序列化的方式得到数据对象，如果缓存文件不存在，则直接进行初始化。当关闭编辑器窗口时，再对数据对象进行序列化并通过文件流写入缓存，代码如下：

```
//第4章/DevelopmentMemo.cs

//缓存文件的路径
private string filePath;
//数据类
[SerializeField] private DevelopmentMemoData data;

private void OnEnable()
{
    //缓存文件的路径
    filePath = Path.GetFullPath(".").Replace("\\", "/")
        + "/Library/DevelopmentMemo.dat";
    //判断是否有缓存文件
    if (File.Exists(filePath))
```

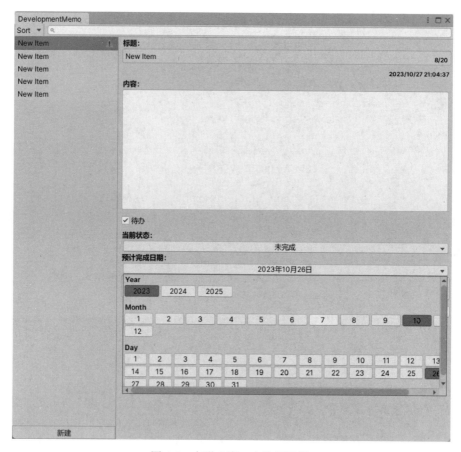

图 4-5 在弹出窗口中选择日期

```
{
    //以文件流的方式打开缓存文件
    using (FileStream fs = File.Open(filePath, FileMode.Open))
    {
        //二进制反序列化
        BinaryFormatter bf = new BinaryFormatter();
        var deserialize = bf.Deserialize(fs);
        if (deserialize != null)
            data = deserialize as DevelopmentMemoData;
        //反序列化失败
        if (data == null)
        {
            //删除无效数据文件
            File.Delete(filePath);
            data = new DevelopmentMemoData();
        }
        //默认按标题排序
        else data.list = data.list.OrderBy(m => m.title).ToList();
```

```
            }
        }
        //当前没有缓存
        else data = new DevelopmentMemoData();
}

private void OnDisable()
{
    try
    {
        //将数据写入缓存
        using (FileStream fs = File.Create(filePath))
        {
            BinaryFormatter bf = new BinaryFormatter();
            bf.Serialize(fs, data);
        }
    }
    catch (Exception ex)
    {
        Debug.LogError(ex);
    }
}
```

当在顶部菜单中单击排序按钮时,通过GenericMenu弹出不同排序方式选项。在排序按钮后面绘制输入框,用于输入检索内容,代码如下:

```
//第 4 章/DevelopmentMemo.cs

//搜索的内容
private string searchContent = string.Empty;

//顶部 GUI
private void OnTopGUI()
{
    GUILayout.BeginHorizontal(EditorStyles.toolbar);
    //排序按钮
    GUI.enabled = data != null && data.list.Count > 0;
    if (GUILayout.Button("Sort", EditorStyles.toolbarDropDown,
        GUILayout.Width(50f)))
    {
        GenericMenu gm = new GenericMenu();
        gm.AddItem(new GUIContent("Title ↓"), false,
            () => data.list = data.list
                .OrderBy(m => m.title).ToList());
        gm.AddItem(new GUIContent("Title ↑"), false,
            () => data.list = data.list
                .OrderByDescending(m => m.title).ToList());
```

```csharp
            gm.AddItem(new GUIContent("Created Time ↓"), false,
                () => data.list = data.list
                    .OrderBy(m => m.createdTime).ToList());
            gm.AddItem(new GUIContent("Created Time ↑"), false,
                () => data.list = data.list
                    .OrderByDescending(m => m.createdTime).ToList());
            gm.ShowAsContext();
        }
        GUI.enabled = true;
        GUILayout.Space(5f);
        //搜索框
        searchContent = GUILayout.TextField(searchContent,
            EditorStyles.toolbarSearchField);
        //当单击鼠标且鼠标位置不在输入框中时取消控件的聚焦
        if (Event.current.type == EventType.MouseDown
            && !GUILayoutUtility.GetLastRect().Contains(
                Event.current.mousePosition))
        {
            GUI.FocusControl(null);
            Repaint();
        }
        GUILayout.EndHorizontal();
    }
```

窗口通过分割线分为左右两部分，通过滚动视图分别绘制备忘录列表、备忘录列表项详情，通过拖动分割线可以调整左右两侧区域的大小，代码如下：

```csharp
//第4章/DevelopmentMemo.cs

//列表的宽度
private float listRectWidth = 280f;
//左右两侧分割线区域
private Rect splitterRect;
//是否正在拖曳分割线
private bool isDragging;
//列表滚动值
private Vector2 listScroll;
//当前选中项
private DevelopmentMemoItem currentItem;
//详情滚动值
private Vector2 detailScroll;
private Rect dateRect;

private void OnGUI()
{
    OnTopGUI();
    OnBodyGUI();
```

```csharp
}
private void OnBodyGUI()
{
    //左侧列表
    GUILayout.BeginHorizontal();
    GUILayout.BeginVertical(GUILayout.Width(listRectWidth));
    OnLeftGUI();
    GUILayout.EndVertical();
    //分割线
    GUILayout.BeginVertical(GUILayout.ExpandHeight(true),
        GUILayout.MaxWidth(5f));
    GUILayout.Box(GUIContent.none, "EyeDropperVerticalLine",
        GUILayout.ExpandHeight(true));
    GUILayout.EndVertical();
    splitterRect = GUILayoutUtility.GetLastRect();
    //右侧详情
    GUILayout.BeginVertical(GUILayout.ExpandWidth(true));
    OnRightGUI();
    GUILayout.EndVertical();
    GUILayout.EndHorizontal();

    if (Event.current != null)
    {
        //光标
        EditorGUIUtility.AddCursorRect(splitterRect,
            MouseCursor.ResizeHorizontal);
        switch (Event.current.rawType)
        {
            //当鼠标被按下时判断是否为分割线的区域
            case EventType.MouseDown:
                isDragging = splitterRect.Contains(
                    Event.current.mousePosition);
                break;
            //在拖曳分割线的过程中根据拖曳偏移量调整左侧列表的宽度
            case EventType.MouseDrag:
                if (isDragging)
                {
                    listRectWidth += Event.current.delta.x;
                    listRectWidth = Mathf.Clamp(listRectWidth,
                        position.width * .3f, position.width * .8f);
                    Repaint();
                }
                break;
            //当鼠标被抬起时结束拖曳
            case EventType.MouseUp:
                if (isDragging)
                    isDragging = false;
```

```csharp
                    break;
            }
        }
    }
    //左侧列表GUI
    private void OnLeftGUI()
    {
        //列表滚动视图
        listScroll = EditorGUILayout.BeginScrollView(listScroll);
        for (int i = 0; i < data.list.Count; i++)
        {
            var item = data.list[i];
            //判断当前项是否符合检索的内容
            if (!item.title.ToLower().Contains(searchContent.ToLower()))
                continue;
            //当前选中项与其他项使用不同的样式
            GUILayout.BeginHorizontal(currentItem == item
                ? "MeTransitionSelectHead"
                : "ProjectBrowserHeaderBgTop");
            GUILayout.Label(item.title);
            if (item.isOverdue)
            {
                GUILayout.FlexibleSpace();
                GUILayout.Label(EditorGUIUtility.IconContent(
                    "console.warnicon.sml"));
            }
            GUILayout.EndHorizontal();
            //当鼠标单击当前项时进行选中操作
            if (Event.current.type == EventType.MouseDown
                && GUILayoutUtility.GetLastRect().Contains(
                    Event.current.mousePosition))
            {
                if (currentItem != item)
                {
                    GUI.FocusControl(null);
                    currentItem = item;
                    Repaint();
                }
            }
        }
        EditorGUILayout.EndScrollView();

        GUILayout.FlexibleSpace();

        GUILayout.Box(GUIContent.none, "EyeDropperHorizontalLine",
            GUILayout.MaxHeight(1f), GUILayout.Width(listRectWidth));
        GUILayout.BeginHorizontal(EditorStyles.toolbar);
```

```csharp
        //新建按钮
        if (GUILayout.Button("新建", EditorStyles.toolbarButton))
        {
            var item = new DevelopmentMemoItem()
            {
                title = "New Item",
                createdTime = DateTime.Now,
                estimateCompleteTime = DateTime.Now.AddDays(3)
            };
            data.list.Add(item);
        }
        GUILayout.EndHorizontal();
    }
    //右侧详情GUI
    private void OnRightGUI()
    {
        //当前未选中任何项
        if (currentItem == null) return;

        detailScroll = EditorGUILayout.BeginScrollView(detailScroll);
        //标题
        GUILayout.Label("标题: ", EditorStyles.boldLabel);
        GUILayout.BeginHorizontal(EditorStyles.helpBox);
        string newTitle = EditorGUILayout.TextField(
            currentItem.title, EditorStyles.label);
        if (newTitle != currentItem.title)
        {
            //长度限制
            if (newTitle.Length > 0 && newTitle.Length <= 20)
                currentItem.title = newTitle;
        }
        GUILayout.FlexibleSpace();
        GUILayout.Label(string.Format("{0}/{1}", currentItem.title.Length,
            20),EditorStyles.miniBoldLabel);
        GUILayout.EndHorizontal();

        //日期
        GUILayout.BeginHorizontal();
        GUILayout.FlexibleSpace();
        GUILayout.Label(currentItem.createdTime.ToString(),
            EditorStyles.miniBoldLabel);
        GUILayout.EndHorizontal();

        //内容
        GUILayout.Label("内容: ", EditorStyles.boldLabel);
        currentItem.content = EditorGUILayout.TextArea(currentItem.content,
            GUILayout.MaxWidth(position.width - listRectWidth - 15f),
```

```
        GUILayout.MinHeight(200f));

EditorGUILayout.Space();
EditorGUI.BeginChangeCheck();
//是否为待办
currentItem.todo = GUILayout.Toggle(currentItem.todo, "待办");
if (EditorGUI.EndChangeCheck())
{
    if (currentItem.todo)
        currentItem.OverdueCal();
    else currentItem.isOverdue = false;

}
EditorGUILayout.Space();

if (currentItem.todo)
{
    //当前状态
    GUILayout.Label("当前状态: ", EditorStyles.boldLabel);
    if (GUILayout.Button(currentItem.isCompleted
        ? "已完成" : "未完成", "DropDownButton"))
    {
        GenericMenu gm = new GenericMenu();
        gm.AddItem(new GUIContent("未完成"), !currentItem.isCompleted,
            () =>
            {
                currentItem.isCompleted = false;
                currentItem.OverdueCal();
            });
        gm.AddItem(new GUIContent("已完成"), currentItem.isCompleted,
            () =>
            {
                currentItem.isCompleted = true;
                currentItem.OverdueCal();
            });
        gm.ShowAsContext();
    }

    if (!currentItem.isCompleted)
    {
        GUILayout.Label("预计完成时间: ",
            EditorStyles.boldLabel);
        GUILayout.BeginHorizontal();
        if (GUILayout.Button(currentItem.estimateCompleteTime
            .ToString("D"), "DropDownButton"))
            PopupWindow.Show(dateRect, new DatePopupWindowContent(
                new Vector2(position.width - listRectWidth - 18f, 200f),
```

```
                currentItem));
        if (Event.current.type == EventType.Repaint)
            dateRect = GUILayoutUtility.GetLastRect();
        GUILayout.EndHorizontal();
        if (currentItem.isOverdue)
            EditorGUILayout.HelpBox("已逾期", MessageType.Warning);
    }
}

//删除
EditorGUILayout.Space();
GUILayout.BeginHorizontal();
GUILayout.FlexibleSpace();
if (GUILayout.Button("删除"))
{
    if (EditorUtility.DisplayDialog("提醒",
        "是否确认删除该项?", "确认", "取消"))
    {
        data.list.Remove(currentItem);
        currentItem = null;
        Repaint();
    }
}
GUILayout.EndHorizontal();
EditorGUILayout.EndScrollView();
}
```

4.1.5　Protobuf 通信协议文件编辑器

在 Socket 网络编程中，假如使用 Protobuf 作为网络通信协议，需要了解 Protobuf 语法规则、编写.proto 文件并通过编译指令将.proto 文件转换为.cs 脚本文件。本节实现一个编辑器工具来使开发者不再需要关注这些语法规则、编译指令，以便更便捷地编辑和修改 proto 文件的内容。

在制作该工具之前，需要先了解 Protobuf 的语法规则，下面是一个.proto 文件的示例：

```
message AvatarProperty
{
    required string userId = 1;
    required float posX = 2;
    required float posY = 3;
    required float posZ = 4;
    required float rotX = 5;
    required float rotY = 6;
    required float rotZ = 7;
    required float speed = 8;
}
```

示例中的 1~8 表示每个字段的标识号，并不是赋值。每个字段都有唯一的标识号，这些标识号是用来在消息的二进制格式中识别各个字段的。[1,15]之内的标识号在编码时会占用一字节。[16,2047]之内的标识号则占用 2 字节，所以应该为那些频繁出现的消息元素保留[1,15]之内的标识号。要为将来有可能添加的频繁出现的标识号预留一些标识号，不可以使用其中的[19000,19999]标识号，Protobuf 协议实现中对这些进行了预留。

类通过 message 声明，后面是类的命名。字段修饰符包含 3 种类型，required 表示不可增加或删除的字段，必须初始化，optional 表示可选字段，可删除，可以不初始化，repeated 表示可重复字段，对应 C#中的 List。Proto 与 C#中字段类型的对应关系见表 4-4。

表 4-4　Proto 与 C#中字段类型的对应关系

.proto	C#	.proto	C#
double	double	fixed32	uint
float	float	fixed64	ulong
int32	int	sfixed32	int
int64	long	sfixed64	long
uint32	uint	bool	bool
uint64	ulong	string	string
sint32	int	bytes	ByteString
sint64	long		

在 Unity 中创建该工具的编辑器窗口类，继承 EditorWindow，使用 MenuItem 定义打开窗口的菜单路径。定义两个枚举，分别表示修饰符的类型和字段类型，代码如下：

```
public class ProtoEditor : EditorWindow
{
    [MenuItem("Example/Multiplayer/Proto Editor")]
    public static void Open()
    {
        GetWindow<ProtoEditor>("Proto Editor").Show();
    }
}
public enum ModifierType
{
    Required,
    Optional,
    Repeated
}
public enum FieldsType
{
    Double,
    Float,
```

```
    Int,
    Long,
    Bool,
    String,
    Custom,
}
```

表示字段的结构中有修饰符、字段类型、字段名称和标识号，在一种通信协议类中可能会包含其他的通信协议类，也就是 Custom 类型。表示类的结构中有类名和存储所有字段的列表，代码如下：

```
//第4章/ProtoEditor.cs
public class Message
{
    //<summary>
    //类名
    //</summary>
    public string name = "New Message";
    //<summary>
    //所有字段
    //</summary>
    public List<Fields> fieldsList = new List<Fields>(0);

    public bool IsValid()
    {
        bool flag = !string.IsNullOrEmpty(name);
        for (int i = 0; i < fieldsList.Count; i++)
        {
            flag &= fieldsList[i].IsValid();
            if (!flag) return false;
            for (int j = 0; j < fieldsList.Count; j++)
            {
                if (i != j)
                {
                    flag &= fieldsList[i].flag != fieldsList[j].flag;
                }
                if (!flag) return false;
            }
        }
        return flag;
    }
}
public class Fields
{
    public ModifierType modifier;
    public FieldsType type;
```

```
    public string typeName;
    public string name;
    public int flag;

    public Fields() { }

    public Fields(int flag)
    {
        modifier = ModifierType.Required;
        type = FieldsType.String;
        name = "FieldsName";
        typeName = "FieldsType";
        this.flag = flag;
    }

    public bool IsValid()
    {
        return type != FieldsType.Custom
            || (type == FieldsType.Custom
            && !string.IsNullOrEmpty(typeName));
    }
}
```

通过输入框控件编辑.proto 文件的名称,然后在滚动视图中列举所有的类,支持类的增删及编辑,代码如下:

```
//第4章/ProtoEditor.cs

//.proto 文件名称
private string fileName;
//存储所有类
private List<Message> messages = new List<Message>();
//滚动值
private Vector2 scroll;
//字段存储折叠状态
private readonly Dictionary<Message, bool> foldoutDic
    = new Dictionary<Message, bool>();

//编辑
private void OnEditGUI()
{
    //编辑.proto 文件名称
    fileName = EditorGUILayout.TextField(".proto File Name", fileName);

    EditorGUILayout.Space();

    //滚动视图
```

```
scroll = GUILayout.BeginScrollView(scroll);
for (int i = 0; i < messages.Count; i++)
{
    var message = messages[i];

    GUILayout.BeginHorizontal();
    foldoutDic[message] = EditorGUILayout.Foldout(
        foldoutDic[message], message.name, true);
    //插入新类
    if (GUILayout.Button("+", GUILayout.Width(20f)))
    {
        Message insertMessage = new Message();
        messages.Insert(i + 1, insertMessage);
        foldoutDic.Add(insertMessage, true);
        Repaint();
        return;
    }
    //删除该类
    if (GUILayout.Button("-", GUILayout.Width(20f)))
    {
        messages.Remove(message);
        foldoutDic.Remove(message);
        Repaint();
        return;
    }
    GUILayout.EndHorizontal();

    //如果折叠栏为打开状态，则绘制具体的字段内容
    if (foldoutDic[message])
    {
        //编辑类名
        message.name = EditorGUILayout.TextField(
            "Name", message.name);
        //字段数量为0，提供按钮创建
        if (message.fieldsList.Count == 0)
        {
            if (GUILayout.Button("New Field"))
            {
                message.fieldsList.Add(new Fields(1));
            }
        }
        else
        {
            for (int j = 0; j < message.fieldsList.Count; j++)
            {
                var item = message.fieldsList[j];
                GUILayout.BeginHorizontal();
```

```csharp
            //修饰符类型
            item.modifier = (ModifierType)EditorGUILayout
                .EnumPopup(item.modifier);
            //字段类型
            item.type = (FieldsType)EditorGUILayout
                .EnumPopup(item.type);
            if (item.type == FieldsType.Custom)
            {
                item.typeName = GUILayout
                    .TextField(item.typeName);
            }
            //编辑字段名
            item.name = EditorGUILayout.TextField(item.name);
            GUILayout.Label("=", GUILayout.Width(15f));
            //分配标识号
            item.flag = EditorGUILayout.IntField(
                item.flag, GUILayout.Width(50f));
            //插入新字段
            if (GUILayout.Button("+", GUILayout.Width(20f)))
            {
                message.fieldsList.Insert(j + 1,
                    new Fields(message.fieldsList.Count + 1));
                Repaint();
                return;
            }
            //删除该字段
            if (GUILayout.Button("-", GUILayout.Width(20f)))
            {
                message.fieldsList.Remove(item);
                Repaint();
                return;
            }
            GUILayout.EndHorizontal();
        }
    }
}
GUILayout.EndScrollView();
}
```

在编辑器窗口底部绘制功能菜单，在功能菜单中支持通信协议类的新建、清空操作，以及数据的序列化与反序列化，以便保存和再次编辑，最终支持生成 .proto 文件和 .bat 文件，代码如下：

```csharp
//第 4 章/ProtoEditor.cs

//JSON 文件的存放路径
```

```csharp
private const string workspacePath
    = "/Metaverse/Scripts/Proto";
private void OnGUI()
{
    OnEditGUI();
    OnBottomMenuGUI();
}
//底部菜单
private void OnBottomMenuGUI()
{
    GUILayout.FlexibleSpace();

    GUILayout.BeginHorizontal();
    //创建新的类
    if (GUILayout.Button("New Message"))
    {
        Message message = new Message();
        messages.Add(message);
        foldoutDic.Add(message, true);
    }
    //清空所有类
    if (GUILayout.Button("Clear Messages"))
    {
        //确认弹窗
        if (EditorUtility.DisplayDialog("Confirm",
            "是否确认清空所有类型？", "确认", "取消"))
        {
            //清空
            messages.Clear();
            foldoutDic.Clear();
            //重新绘制
            Repaint();
        }
    }
    GUILayout.EndHorizontal();

    GUILayout.BeginHorizontal();
    //导出 JSON
    if (GUILayout.Button("Export JSON File"))
    {
        if (!ContentIsValid())
        {
            EditorUtility.DisplayDialog("Error",
                "请按以下内容逐项检查：\r\n" +
                "1.proto File Name 是否为空\r\n" +
                "2.message 类名是否为空\r\n" +
                "3.当字段类型为自定义时是否填写了类型名称\r\n" +
```

```csharp
            "4.标识号是否唯一", "OK");
    }
    else
    {
        //文件夹路径
        string dirPath = string.Format("{0}{1}/json",
            Application.dataPath, workspacePath);
        //如果文件夹不存在，则创建
        if (!Directory.Exists(dirPath))
            Directory.CreateDirectory(dirPath);
        //JSON 文件路径
        string filePath = dirPath + "/" + fileName + ".json";
        if (EditorUtility.DisplayDialog("Confirm",
            "是否保存当前编辑内容到" + filePath, "确认", "取消"))
        {
            //序列化
            string json = JsonMapper.ToJson(messages);
            //写入
            File.WriteAllText(filePath, json);
            //刷新
            AssetDatabase.Refresh();
        }
    }
}
//导入JSON
if (GUILayout.Button("Import JSON File"))
{
    //选择JSON文件路径
    string filePath = EditorUtility.OpenFilePanel("Import JSON File",
        string.Format("{0}{1}/json",
            Application.dataPath, workspacePath), "json");
    //判断路径有效性
    if (File.Exists(filePath))
    {
        //读取 JSON 内容
        string json = File.ReadAllText(filePath);
        //清空
        messages.Clear();
        foldoutDic.Clear();
        //反序列化
        messages = JsonMapper.ToObject<List<Message>>(json);
        //填充字典
        for (int i = 0; i < messages.Count; i++)
        {
            foldoutDic.Add(messages[i], true);
        }
        //文件名称
```

```csharp
            FileInfo fileInfo = new FileInfo(filePath);
            fileName = fileInfo.Name.Replace(".json", "");
            //重新绘制
            Repaint();
            return;
        }
    }
    GUILayout.EndHorizontal();

    //生成.proto文件
    if (GUILayout.Button("Generate Proto File"))
    {
        if (!ContentIsValid())
        {
            EditorUtility.DisplayDialog("Error",
                "请按以下内容逐项检查: \r\n" +
                "1.proto File Name 是否为空\r\n" +
                "2.message 类名是否为空\r\n" +
                "3.当字段类型为自定义时是否填写了类型名称\r\n" +
                "4.标识号是否唯一", "OK");
        }
        else
        {
            string protoFilePath = EditorUtility.SaveFilePanel(
                "Generate Proto File", new DirectoryInfo(Application.dataPath)
                    .Parent.FullName + "/protogen/Proto", fileName, "proto");
            if (!string.IsNullOrEmpty(protoFilePath))
            {
                StringBuilder protoContent = new StringBuilder();
                for (int i = 0; i < messages.Count; i++)
                {
                    var message = messages[i];
                    StringBuilder sb = new StringBuilder();
                    sb.Append("message " + message.name + "\r\n" + "{\r\n");
                    for (int n = 0; n < message.fieldsList.Count; n++)
                    {
                        var field = message.fieldsList[n];
                        //缩进
                        sb.Append("    ");
                        //修饰符
                        sb.Append(field.modifier.ToString().ToLower());
                        //空格
                        sb.Append(" ");
                        //如果是自定义类型，则拼接 typeName
                        switch (field.type)
                        {
                            case
```

```
                            FieldsType.Int:
                                sb.Append("int32");
                                break;
                            case
                            FieldsType.Long:
                                sb.Append("int64");
                                break;
                            case
                            FieldsType.Custom:
                                sb.Append(field.typeName);
                                break;
                            default:
                                sb.Append(field.type.ToString().ToLower());
                                break;
                        }
                        //空格
                        sb.Append(" ");
                        //字段名
                        sb.Append(field.name);
                        //等号
                        sb.Append(" = ");
                        //标识号
                        sb.Append(field.flag);
                        //分号及换行符
                        sb.Append(";\r\n");
                    }
                    sb.Append("}\r\n");
                    protoContent.Append(sb.ToString());
                }
                //写入文件
                File.WriteAllText(protoFilePath, protoContent.ToString());
                //刷新(假设路径在工程内,可以避免手动刷新才能看到)
                AssetDatabase.Refresh();
                //打开该文件夹
                FileInfo fileInfo = new FileInfo(protoFilePath);
                Process.Start(fileInfo.Directory.FullName);
            }
        }
    }

    //创建.bat 文件
    if (GUILayout.Button("Create .bat"))
    {
        //选择路径(protogen.exe 所在的文件夹路径)
        string rootPath = EditorUtility.OpenFolderPanel(
            "Create .bat file (protogen.exe 所在的文件夹)",
            new DirectoryInfo(Application.dataPath).Parent.FullName
```

```csharp
                + "/protogen", string.Empty);
        //取消
        if (string.IsNullOrEmpty(rootPath)) return;
        //protogen.exe 文件路径
        string protogenPath = rootPath + "/protogen.exe";
        //不是protogen.exe所在的文件路径
        if (!File.Exists(protogenPath))
        {
            EditorUtility.DisplayDialog("Error",
                "请选择protogen.exe所在的文件路径", "OK");
        }
        else
        {
            string protoPath = rootPath + "/Proto";
            DirectoryInfo di = new DirectoryInfo(protoPath);
            //获取所有.proto文件信息
            FileInfo[] protos = di.GetFiles("*.proto");
            //使用StringBuilder拼接字符串
            StringBuilder sb = new StringBuilder();
            //遍历
            for (int i = 0; i < protos.Length; i++)
            {
                string proto = protos[i].Name;
                //拼接编译指令
                sb.Append(rootPath + @"/protogen.exe -i:Proto\"
                    + proto + @" -o:cs\"
                    + proto.Replace(".proto", ".cs") + "\r\n");
            }
            sb.Append("pause");

            //生成.bat文件
            string batPath = $"{rootPath}/run.bat";
            File.WriteAllText(batPath, sb.ToString());
            //打开该文件夹
            Process.Start(rootPath);
        }
    }
}
//编辑的内容是否有效
private bool ContentIsValid()
{
    bool flag = !string.IsNullOrEmpty(fileName);
    flag &= messages.Count > 0;
    for (int i = 0; i < messages.Count; i++)
    {
        flag &= messages[i].IsValid();
        if (!flag) break;
```

```
        }
        return flag;
}
```

结果如图 4-6 所示。

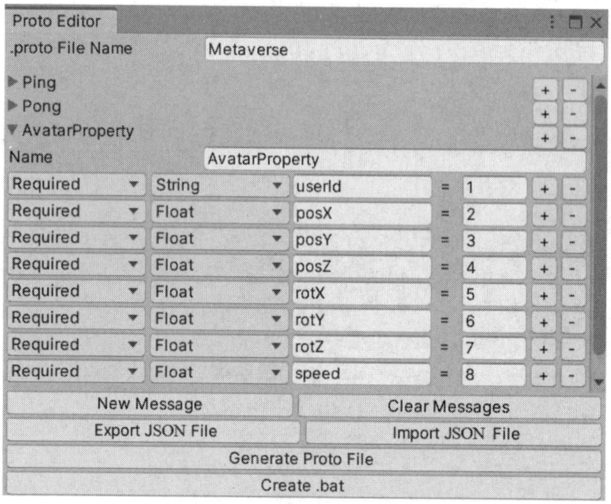

图 4-6　Protobuf 通信协议文件编辑器

运行工具生成的 .bat 文件，即可将 proto 文件夹下的通信协议文件编译为 C# 脚本文件并存储到 cs 文件夹中，如图 4-7 所示。

图 4-7　protogen 编译通信协议

4.1.6 ScriptableWizard

ScriptableWizard 是 EditorWindow 的派生类，该类型的窗口称为向导型编辑器窗口。在向导型编辑器窗口类中可序列化属性会自动地在窗口中提供交互修改的控件，不需要再调用相关创建控件的方法。

打开该类型的窗口需要调用 ScriptableWizard 类中的静态方法 DisplayWizard()，此方法的代码如下，参数 title 表示窗口的标题，createButtonName 表示 create 按钮上显示的文本，otherButtonName 表示 other 按钮上显示的文本。

```
public static T DisplayWizard (string title);
public static T DisplayWizard (string title, string createButtonName);
public static T DisplayWizard (string title, string createButtonName, string otherButtonName);
```

向导型编辑器窗口在窗口的右下角区域有两个在内部自动实现的按钮，分别是 create 和 other 按钮，这两个按钮的单击事件分别是在向导型编辑器窗口类中实现的 OnWizardCreate() 和 OnWizardOtherButton() 方法。查看 ScriptableWizard 类内部的 OnGUI() 方法，可以看到两个按钮的单击事件是通过反射的形式被调用的，代码如下：

```
private void OnGUI()
{
    EditorGUIUtility.labelWidth = 150f;
    GUILayout.Label(m_HelpString, EditorStyles.wordWrappedLabel,
        GUILayout.ExpandHeight(expand: true));
    m_ScrollPosition = EditorGUILayout.BeginVerticalScrollView(
        m_ScrollPosition, false, GUI.skin.verticalScrollbar, "OL Box");
    GUIUtility.GetControlID(645789, FocusType.Passive);
    bool flag = DrawWizardGUI();
    EditorGUILayout.EndScrollView();
    GUILayout.BeginVertical();
    if (m_ErrorString != string.Empty)
    {
        GUILayout.Label(m_ErrorString, Styles.errorText,
            GUILayout.MinHeight(32f));
    }
    else
    {
        GUILayout.Label(string.Empty, GUILayout.MinHeight(32f));
    }

    GUILayout.FlexibleSpace();
    GUILayout.BeginHorizontal();
    GUILayout.FlexibleSpace();
    GUI.enabled = m_IsValid;
    if (m_OtherButton != "" && GUILayout.Button(
```

```csharp
        m_OtherButton, GUILayout.MinWidth(100f)))
{
    MethodInfo method = GetType().GetMethod("OnWizardOtherButton",
        BindingFlags.FlattenHierarchy | BindingFlags.Instance
        | BindingFlags.NonPublic | BindingFlags.Public);
    if ((object)method != null)
    {
        method.Invoke(this, null);
        GUIUtility.ExitGUI();
    }
    else
    {
        Debug.LogError("OnWizardOtherButton has not " +
            "been implemented in script");
    }
}

if (m_CreateButton != "" && GUILayout.Button(
    m_CreateButton, GUILayout.MinWidth(100f)))
{
    MethodInfo method2 = GetType().GetMethod("OnWizardCreate",
        BindingFlags.FlattenHierarchy | BindingFlags.Instance
        | BindingFlags.NonPublic | BindingFlags.Public);
    if ((object)method2 != null)
    {
        method2.Invoke(this, null);
    }
    else
    {
        Debug.LogError("OnWizardCreate has not " +
            "been implemented in script");
    }

    Close();
    GUIUtility.ExitGUI();
}

GUI.enabled = true;
GUILayout.EndHorizontal();
GUILayout.EndVertical();
if (flag)
{
    InvokeWizardUpdate();
}

GUILayout.Space(8f);
}
```

假设有一个驱动人物角色移动的脚本组件 AvatarController，人物驱动是通过 CharacterController 组件实现的，因此在场景中创建一个玩家的人物角色时，需要将人物模型拖入场景中，然后为其添加 CharacterController 和 AvatarController 组件，因为人物角色还会有人物动画，所以还需要添加一个 Animator 组件。

这个创建的过程可以通过向导型编辑器窗口实现，在 OnWizardCreate() 方法中获取用户选择的人物模型，将其实例化到场景中，然后为其添加相应的组件完成人物角色的创建，代码如下：

```csharp
//第4章/AvatarCreateScriptableWizard.cs

using UnityEngine;
using UnityEditor;
using UnityEditor.Animations;

//<summary>
//人物创建向导窗口
//</summary>
public class AvatarCreateScriptableWizard : ScriptableWizard
{
    [MenuItem("Example/Avatar Creator")]
    public static void Open()
    {
        DisplayWizard<AvatarCreateScriptableWizard>(
            "Avatar Creator", "创建人物角色");
    }

    public AnimatorController animatorController;
    public bool applyRootMotion;

    private void OnWizardCreate()
    {
        //未选中任何对象
        if (Selection.activeGameObject == null)
            return;
        //实例化
        GameObject instance = Instantiate(Selection.activeGameObject);
        instance.transform.localPosition = Vector3.zero;
        instance.transform.localRotation = Quaternion.identity;
        instance.transform.localScale = Vector3.one;
        //为人物添加角色控制器组件
        instance.AddComponent<CharacterController>();
        //为人物添加自定义的人物驱动组件
        instance.AddComponent<AvatarController>();
        //为人物添加动画组件
        Animator animator = instance.GetComponent<Animator>();
```

```
            if (animator == null) animator = instance.AddComponent<Animator>();
            animator.runtimeAnimatorController = animatorController;
            animator.applyRootMotion = applyRootMotion;
        }
    }
```

如图 4-8 所示,在 Project 窗口中选择一个人物模型,单击 Avatar Creator 窗口中创建人物角色的按钮后,该窗口在场景中创建了一个人物角色实例,该实例自动地添加了动画、角色控制器等相应的组件。

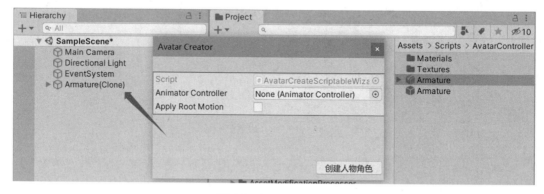

图 4-8 Avatar Creator

4.2 如何扩展默认的编辑器窗口

4.2.1 扩展 Hierarchy 窗口

拓展层级窗口通过 EditorApplication 类中的 hierarchyWindowItemOnGUI 委托方法实现,其中 instanceID 参数表示的是游戏物体的实例 ID,有了实例 ID,可以通过 EditorUtility 类中的 InstanceIDToObject() 方法获取对应的游戏物体。第 2 个参数 selectionRect 表示的是在层级窗口绘制该游戏物体的矩形区域。

例如想要在矩形区域后面绘制出该游戏物体挂载的组件的缩略图,首先需要获取组件的类型,然后通过 AssetPreview 类中的 GetMiniTypeThumbnail() 方法获取该类型的缩略图,代码如下:

```
//第 4 章/CustomHierarchyWindow.cs

using UnityEngine;
using UnityEditor;

public class CustomHierarchyWindow
{
    [InitializeOnLoadMethod]
    static void InitializeOnLoad()
```

```csharp
{
    EditorApplication.hierarchyWindowItemOnGUI
        -= OnHierarchyWindowItemGUI;
    EditorApplication.hierarchyWindowItemOnGUI
        += OnHierarchyWindowItemGUI;
}

private static void OnHierarchyWindowItemGUI(
    int instanceID, Rect selectionRect)
{
    GameObject go = EditorUtility
        .InstanceIDToObject(instanceID) as GameObject;
    if (go == null) return;
    Component[] components = go.GetComponents<Component>();
    for (int i = 0; i < components.Length; i++)
    {
        Component component = components[i];
        if (component == null) continue;
        Texture texture = AssetPreview.GetMiniTypeThumbnail(
            component.GetType()) ??
            AssetPreview.GetMiniThumbnail(component);
        if (texture == null) continue;
        Rect rect = selectionRect;
        rect.x += selectionRect.width - (i + 1) * 20f;
        rect.width = 20f;
        GUI.Label(rect, new GUIContent(texture,
            component.GetType().Name));
    }
}
```

结果如图 4-9 所示。

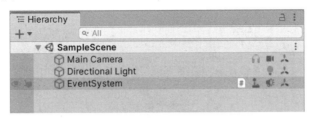

图 4-9　扩展 Hierarchy 窗口

4.2.2　扩展 Project 窗口

扩展 Project 窗口通过 EditorApplication 类中的 projectWindowItemOnGUI 委托方法实现，其中 guid 参数表示的是资产的全局唯一标识符，在 Unity 中，每个资产都有一个对应的 guid 来标识它，AssetDatabase 类中的 GUIDToAssetPath() 方法可以通过 guid 获取对应的资产路径。

第 2 个参数 selectionRect 表示的是在 Project 窗口绘制该资产时的矩形区域。例如，想要在矩形区域后面绘制出该资产的依赖项的数量，代码如下：

```csharp
//第 4 章/CustomProjectWindow.cs

using UnityEngine;
using UnityEditor;

public class CustomProjectWindow
{
    [InitializeOnLoadMethod]
    static void InitializeOnLoad()
    {
        EditorApplication.projectWindowItemOnGUI
            -= OnProjectWindowItemGUI;
        EditorApplication.projectWindowItemOnGUI
            += OnProjectWindowItemGUI;
    }

    private static void OnProjectWindowItemGUI(
        string guid, Rect selectionRect)
    {
        string assetPath = AssetDatabase.GUIDToAssetPath(guid);
        string[] dependencies = AssetDatabase.GetDependencies(assetPath);
        if (dependencies.Length == 0) return;
        Rect rect = selectionRect;
        rect.x += selectionRect.width - 20f;
        rect.width = 20f;
        GUI.Label(rect, dependencies.Length.ToString());
    }
}
```

结果如图 4-10 所示。

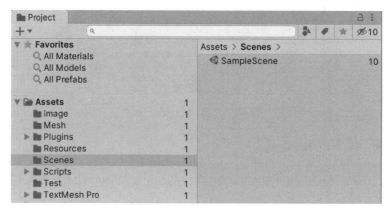

图 4-10　扩展 Project 窗口

4.3 Game 窗口中的 GUI

OnGUI()是 MonoBehaviour 中的一个重要的回调方法，在项目开发中，除了可以使用控制台输出日志作为调试手段外，可以使用该方法在 Game 窗口中绘制出相关的变量值，便于查看及调试。该方法也可以在项目初期没有 UI 界面时，用于绘制按钮等控件，以此充当临时 UI。

在开发工作中，有时会希望缩放 TimeScale 来调试某一段程序的功能，这时便可以使用 OnGUI()方法，在 Game 窗口中绘制一些按钮，用于控制时间缩放值，代码如下：

```
//第 4 章/MonoBehaviourOnGUIExample.cs
using UnityEngine;
public class MonoBehaviourOnGUIExample : MonoBehaviour
{
    private void OnGUI()
    {
        TimeScaleExample();
    }
    private void TimeScaleExample()
    {
        GUILayout.BeginHorizontal();
        GUILayout.Label("TimeScale",
            GUILayout.Width(100f));
        if (GUILayout.Button("0.1f",
            GUILayout.Width(100f), GUILayout.Height(50f)))
            Time.timeScale = 0.1f;
        if (GUILayout.Button("0.25f",
            GUILayout.Width(100f), GUILayout.Height(50f)))
            Time.timeScale = 0.25f;
        if (GUILayout.Button("0.5f",
            GUILayout.Width(100f), GUILayout.Height(50f)))
            Time.timeScale = 0.5f;
        if (GUILayout.Button("1",
            GUILayout.Width(100f), GUILayout.Height(50f)))
            Time.timeScale = 1f;
        GUILayout.EndHorizontal();
    }
}
```

运行程序后，Game 窗口左上角如图 4-11 所示。

GUI 类中的 Window()方法可以在 Game 窗口中创建一个浮窗，该窗口浮动在 GUI 控件的上方，通过单击可以获得焦点，可以定义是否允许用户拖曳，方法具有多个重载，代

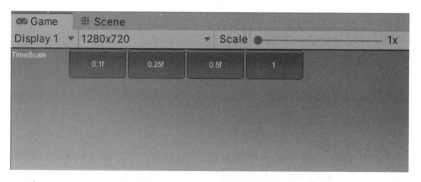

图 4-11 Game 窗口中的 GUI

码如下：

```
public static Rect Window (int id, Rect clientRect, GUI.WindowFunction func,
string text);
public static Rect Window (int id, Rect clientRect, GUI.WindowFunction func,
Texture image);
public static Rect Window (int id, Rect clientRect, GUI.WindowFunction func,
GUIContent content);
public static Rect Window (int id, Rect clientRect, GUI.WindowFunction func,
string text, GUIStyle style);
public static Rect Window (int id, Rect clientRect, GUI.WindowFunction func,
Texture image, GUIStyle style);
public static Rect Window (int id, Rect clientRect, GUI.WindowFunction func,
GUIContent title, GUIStyle style);
```

参数 id 表示窗口的编号，可以是任意值，只需保证其唯一性；clientRect 是表示窗口位置和大小的矩形区域；func 表示定义窗口中内容的方法；text 表示要在窗口内绘制的文本；image 表示要在窗口中绘制的图像；content 表示要在窗口中呈现的 GUIContent；style 表示窗口使用的样式；title 表示在窗口标题栏中绘制的文本。

示例代码如下：

```
//第 4 章/MonoBehaviourOnGUIExample.cs

private Rect windowRect = new Rect(0f, 0f, 300f, 100f);

private void OnGUI()
{
    windowRect = GUI.Window(0, windowRect,
        OnWindowGUI, "Example Window");
}
private void OnWindowGUI(int windowId)
{
    GUILayout.Label("Hello World.");
}
```

运行程序后，Game 窗口左上角如图 4-12 所示。

GUI 类中的 DragWindow()方法可以使该窗口可拖动，调用该方法时需要传入表示窗口可拖动区域的 Rect 类型参数，图 4-12 所示的窗口宽度为 300，高度为 100，假如想要使该窗口标题部分区域可以拖曳，那么参数可以取值为(0f, 0f, 300f, 20f)，代码如下：

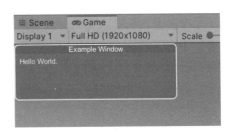

图 4-12　GUI Window

```
//第 4 章/MonoBehaviourOnGUIExample.cs

private Rect windowRect = new Rect(0f, 0f, 300f, 100f);

private void OnGUI()
{
    windowRect = GUI.Window(0, windowRect,
        OnWindowGUI, "Example Window");
}
private void OnWindowGUI(int windowId)
{
    GUI.DragWindow(new Rect(0f, 0f, 300f, 20f));
    GUILayout.Label("Hello World.");
}
```

4.3.1　运行时控制台窗口

本节使用 GUI.Window()方法在 Game 窗口中创建一个控制台窗口，在窗口中可以查看程序的所有运行日志，便于在非 Editor 环境中调试程序，如图 4-13 所示。

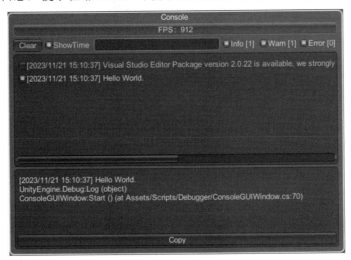

图 4-13　运行时控制台窗口

实现该窗口工具首先需要定义日志项的数据结构，包括日志的类型、时间、堆栈信息、摘要、详情等字段，以及控制台工作的类型，代码如下：

```csharp
//第4章/ConsoleGUIWindow.cs

public class ConsoleItem
{
    public LogType type;
    public DateTime time;
    public string message;
    public string stackTrace;
    public string brief;
    public string detail;

    public ConsoleItem(DateTime time, LogType type,
        string message, string stackTrace)
    {
        this.type = type;
        this.time = time;
        this.message = message;
        this.stackTrace = stackTrace;
        brief = string.Format("[{0}] {1}",
            time, message);
        detail = string.Format("[{0}] {1}\r\n{2}",
            time, message, stackTrace);
    }
}

public enum WorkingType
{
    //<summary>
    //始终打开
    //</summary>
    ALWAYS_OPEN,
    //<summary>
    //仅在 Development Build 模式下打开
    //</summary>
    ONLY_OPEN_WHEN_DEVELOPMENT_BUILD,
    //<summary>
    //仅在 Editor 中打开
    //</summary>
    ONLY_OPEN_IN_EDITOR,
    //<summary>
    //始终关闭
    //</summary>
    ALWAYS_CLOSE
}
```

第 2 种工作类型是指程序构建时在 Build Settings 窗口中勾选 Development Build 设置选项，如图 4-14 所示，该设置可以通过 Debug 类中的静态变量 isDebugBuild 获取。

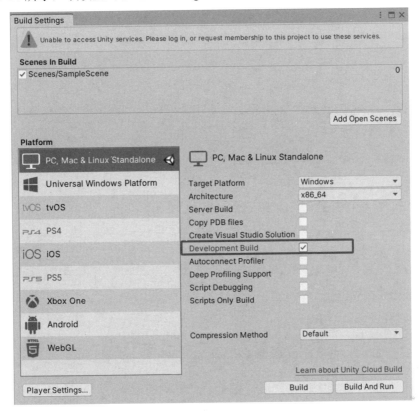

图 4-14 Development Build

窗口默认以收起的状态显示，仅展示当前的 FPS 数据，在单击 FPS 数据处的按钮时展开窗口，当再次单击时收起窗口。

通过 DrawWindow() 方法使窗口可拖动，但是需要限制取值范围来限制窗口可拖动的区域，防止窗口被拖到屏幕外。

日志打印事件通过 Application.logMessageReceived 进行注册,在收到新的日志消息时,创建新的日志项数据，将其存储于列表中，代码如下：

```
//第4章/ConsoleGUIWindow.cs

[SerializeField]
private WorkingType workingType = WorkingType.ALWAYS_OPEN;
private Rect expandRect;
private Rect retractRect;
private Rect dragableRect;
private bool isExpand;
private int fps;
```

```csharp
private float lastShowFPSTime;
private Color fpsColor = Color.white;
//日志列表
private readonly List<ConsoleItem> logs = new List<ConsoleItem>();

private void Start()
{
    switch (workingType)
    {
        case WorkingType.ALWAYS_OPEN:
            enabled = true;
            break;
        case WorkingType.ONLY_OPEN_WHEN_DEVELOPMENT_BUILD:
            enabled = Debug.isDebugBuild;
            break;
        case WorkingType.ONLY_OPEN_IN_EDITOR:
            enabled = Application.isEditor;
            break;
        case WorkingType.ALWAYS_CLOSE:
            enabled = false;
            break;
    }
    if (!enabled) return;

    expandRect = new Rect(Screen.width * .7f, 0f,
        Screen.width * .3f, Screen.height * .5f);
    retractRect = new Rect(Screen.width - 100f, 0f, 100f, 60f);
    dragableRect = new Rect(0, 0, Screen.width * .3f, 20f);
    //事件注册
    Application.logMessageReceived += OnLogMessageReceived;
}
private void OnDestroy()
{
    Application.logMessageReceived -= OnLogMessageReceived;
}

private void OnLogMessageReceived(string condition,
    string stackTrace, LogType logType)
{
    var item = new ConsoleItem(DateTime.Now,
        logType, condition, stackTrace);
    if (logType == LogType.Log) infoCount++;
    else if (logType == LogType.Warning) warnCount++;
    else errorCount++;
    logs.Add(item);
    if (logs.Count > maxCacheCount)
    {
```

```csharp
            logs.RemoveAt(0);
        }
    }

    private void OnGUI()
    {
        if (isExpand)
        {
            expandRect = GUI.Window(0, expandRect, OnExpandGUI, "DebugGER");
            //限制范围
            expandRect.x = Mathf.Clamp(expandRect.x, 0, Screen.width * .7f);
            expandRect.y = Mathf.Clamp(expandRect.y, 0, Screen.height * .5f);
            dragableRect = new Rect(0, 0, Screen.width * .3f, 20f);
        }
        else
        {
            retractRect = GUI.Window(0, retractRect, OnRetractGUI, "DebugGER");
            //限制范围
            retractRect.x = Mathf.Clamp(retractRect.x, 0, Screen.width - 100f);
            retractRect.y = Mathf.Clamp(retractRect.y, 0, Screen.height - 60f);
            dragableRect = new Rect(0, 0, 100f, 20f);
        }
        //FPS 计算
        if (Time.realtimeSinceStartup - lastShowFPSTime >= 1)
        {
            fps = Mathf.RoundToInt(1f / Time.deltaTime);
            lastShowFPSTime = Time.realtimeSinceStartup;
            fpsColor = errorCount > 0 ? Color.red
                : warnCount > 0 ? Color.yellow : Color.white;
        }
    }
    //窗口为收起状态
    private void OnRetractGUI(int windowId)
    {
        GUI.DragWindow(dragableRect);
        GUI.contentColor = fpsColor;
        if (GUILayout.Button(string.Format("FPS: {0}", fps),
            GUILayout.Height(30f)))
            isExpand = true;
        GUI.contentColor = Color.white;
    }
    //窗口为展开状态
    private void OnExpandGUI(int windowId)
    {
        GUI.DragWindow(dragableRect);
        GUI.contentColor = fpsColor;
        if (GUILayout.Button(string.Format("FPS: {0}", fps),
```

```
            GUILayout.Height(20f)))
        isExpand = false;
}
```

当窗口为展开状态时,在滚动视图中展示日志列表,列表中显示日志的简要信息,当选中日志项时,可以在下方的窗口中查看该日志的详细堆栈信息,单击 Copy 按钮可将详细信息复制到系统粘贴板中,代码如下:

```
//第 4 章/ConsoleGUIWindow.cs

//列表滚动视图
private Vector2 listScroll;
//详情滚动视图
private Vector2 detailScroll;
//普通日志数量
private int infoCount;
//告警日志数量
private int warnCount;
//错误日志数量
private int errorCount;
//是否显示普通日志
[SerializeField] private bool showInfo = true;
//是否显示告警日志
[SerializeField] private bool showWarn = true;
//是否显示错误日志
[SerializeField] private bool showError = true;
//当前选中的日志项
private ConsoleItem currentSelected;
//是否显示日志时间
[SerializeField] private bool showTime = true;
//最大缓存数量
[SerializeField] private int maxCacheCount = 100;
//检索内容
private string searchContent;

//窗口为展开状态
private void OnExpandGUI(int windowId)
{
    GUI.DragWindow(dragableRect);
    GUI.contentColor = fpsColor;
    if (GUILayout.Button(string.Format("FPS: {0}", fps),
        GUILayout.Height(20f)))
            isExpand = false;
    OnTopGUI();
    OnListGUI();
    OnDetailGUI();
}
```

```csharp
private void OnTopGUI()
{
    GUILayout.BeginHorizontal();
    //清空所有日志
    if (GUILayout.Button("Clear", GUILayout.Width(50f)))
    {
        logs.Clear();
        infoCount = 0;
        warnCount = 0;
        errorCount = 0;
        currentSelected = null;
    }
    //是否显示日志时间
    showTime = GUILayout.Toggle(showTime, "ShowTime",
        GUILayout.Width(80f));

    //检索输入框
    searchContent = GUILayout.TextField(searchContent,
        GUILayout.ExpandWidth(true));

    GUI.contentColor = showInfo ? Color.white : Color.grey;
    showInfo = GUILayout.Toggle(showInfo, string.Format(
        "Info [{0}]", infoCount), GUILayout.Width(60f));
    GUI.contentColor = showWarn ? Color.white : Color.grey;
    showWarn = GUILayout.Toggle(showWarn, string.Format(
        "Warn [{0}]", warnCount), GUILayout.Width(65f));
    GUI.contentColor = showError ? Color.white : Color.grey;
    showError = GUILayout.Toggle(showError, string.Format(
        "Error [{0}]", errorCount), GUILayout.Width(65f));
    GUI.contentColor = Color.white;
    GUILayout.EndHorizontal();
}
private void OnListGUI()
{
    GUILayout.BeginVertical("Box",
        GUILayout.Height(Screen.height * .3f));
    //滚动视图
    listScroll = GUILayout.BeginScrollView(listScroll);
    for (int i = logs.Count - 1; i >= 0; i--)
    {
        var temp = logs[i];
        //是否符合检索内容
        if (!string.IsNullOrEmpty(searchContent) && !temp.message
            .ToLower().Contains(searchContent.ToLower())) continue;
        bool show = false;
        switch (temp.type)
        {
```

```csharp
                case LogType.Log:
                    if (showInfo) show = true;
                    break;
                case LogType.Warning:
                    if (showWarn) show = true;
                    GUI.contentColor = Color.yellow;
                    break;
                case LogType.Error:
                case LogType.Assert:
                case LogType.Exception:
                    if (showError) show = true;
                    GUI.contentColor = Color.red;
                    break;
            }
            if (show)
            {
                if (GUILayout.Toggle(currentSelected == temp,
                    showTime ? temp.brief : temp.message))
                    currentSelected = temp;
            }
            GUI.contentColor = Color.white;
        }
        GUILayout.EndScrollView();
        GUILayout.EndVertical();
    }
    private void OnDetailGUI()
    {
        GUILayout.BeginVertical("Box", GUILayout.ExpandHeight(true));
        detailScroll = GUILayout.BeginScrollView(detailScroll);
        if (currentSelected != null)
        {
            GUILayout.Label(currentSelected.detail);
        }
        GUILayout.EndScrollView();
        GUILayout.FlexibleSpace();
        GUI.enabled = currentSelected != null;
        //单击按钮时将日志详情复制到系统粘贴板中
        if (GUILayout.Button("Copy", GUILayout.Height(20f)))
        {
            GUIUtility.systemCopyBuffer = currentSelected.detail;
        }
        GUILayout.EndVertical();
    }
```

4.3.2 运行时层级窗口

4.3.1 节介绍了控制台窗口工具的示例，除此之外，开发者可以使用 GUI 窗口制作任何有助于在非编辑器环境中调试程序的工具。例如在一个新的 GUI 窗口中绘制物体的层级关系，类似于编辑器中的 Hierarchy 窗口，如图 4-15 所示。

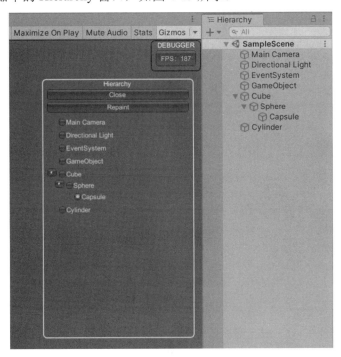

图 4-15　Hierarchy GUI Window

实现该窗口需要收集物体的层级关系，每个物体对应一条层级项数据，数据包括该物体的子项数据、是否展开等字段信息，并且提供绘制该层级节点的方法，代码如下：

```
//第 4 章/HierarchyGUIWindowItem.cs

using UnityEngine;
using System.Collections.Generic;

public class HierarchyGUIWindowItem
{
    private readonly Transform transform;
    private readonly HierarchyGUIWindow window;
    public readonly List<HierarchyGUIWindowItem> childrens;
    private bool expand;
    private int level;

    public HierarchyGUIWindowItem(Transform transform,
```

```csharp
        HierarchyGUIWindow window)
{
    this.transform = transform;
    this.window = window;
    childrens = new List<HierarchyGUIWindowItem>();
    GetParent(transform);
}
private void GetParent(Transform transform)
{
    Transform parent = transform.parent;
    if (parent != null)
    {
        level++;
        GetParent(parent);
    }
}

public void Draw()
{
    if (transform == null) return;

    GUILayout.BeginHorizontal();
    GUILayout.Space(15f * level);
    if (transform.childCount > 0)
    {
        if (GUILayout.Button(expand ? "▼" : "▶",
            GUILayout.Width(17.5f), GUILayout.Height(15f)))
        {
            expand = !expand;
        }
    }
    else
    {
        GUILayout.Label(GUIContent.none, GUILayout.Width(17.5f));
    }
    if (GUILayout.Toggle(window.currentSelected
        == transform.gameObject, transform.name))
    {
        window.currentSelected = transform.gameObject;
    }
    GUILayout.EndHorizontal();
    if (expand)
    {
        for (int i = 0; i < childrens.Count; i++)
        {
            childrens[i].Draw();
        }
```

当绘制 GUI 窗口时，注意要保证窗口 ID 的唯一性。由于日志工具已经使用 0 作为窗口 ID，所以此处使用 1 作为层级窗口的 ID。

与控制台窗口一致，层级窗口分为展开、收起两种状态，在展开状态时绘制物体的层级关系，并且可以通过 Toggle 控件选择物体，代码如下：

```csharp
//第 4 章/HierarchyGUIWindow.cs
using UnityEngine;
using System.Collections.Generic;
using UnityEngine.SceneManagement;
public class HierarchyGUIWindow : MonoBehaviour
{
    private Rect expandRect;
    private Rect retractRect;
    private Rect dragableRect;
    private bool isExpand;
    //层级列表
    private List<HierarchyGUIWindowItem> list;
    //滚动视图的滚动值
    private Vector2 scroll;
    //当前选中的物体
    [HideInInspector] public GameObject currentSelected;

    private void OnEnable()
    {
        list = new List<HierarchyGUIWindowItem>();
        CollectRoots();

        expandRect = new Rect(0f, 0f, 300f, 500f);
        retractRect = new Rect(0f, 0f, 100f, 60f);
        dragableRect = new Rect(0f, 0f, 300f, 20f);
    }
    private void OnDisable()
    {
        list.Clear();
        list = null;
    }

    //收集根级物体
    private void CollectRoots()
    {
        list.Clear();
```

```csharp
        for (int i = 0; i < SceneManager.sceneCount; i++)
        {
            var scene = SceneManager.GetSceneAt(i);
            if (!scene.isLoaded) continue;

            var roots = scene.GetRootGameObjects();
            for (int j = 0; j < roots.Length; j++)
            {
                CollectChildrens(roots[j].transform, list);
            }
        }
    }
    //收集子物体
    private void CollectChildrens(Transform transform,
        List<HierarchyGUIWindowItem> list)
    {
        var item = new HierarchyGUIWindowItem(transform, this);
        list.Add(item);
        for (int i = 0; i < transform.childCount; i++)
        {
            CollectChildrens(transform.GetChild(i),
                item.childrens);
        }
    }

    private void OnGUI()
    {
        if (isExpand)
        {
            expandRect = GUI.Window(1, expandRect,
                OnExpandGUI, "Hierarchy");
            //限制窗口拖动范围
            expandRect.x = Mathf.Clamp(expandRect.x,
                0f, Screen.width - 300f);
            expandRect.y = Mathf.Clamp(expandRect.y,
                0f, Screen.height - 500f);
            dragableRect = new Rect(0f, 0f, 300f, 20f);
        }
        else
        {
            retractRect = GUI.Window(1, retractRect,
                OnRetractGUI, "Hierarchy");
            //限制窗口拖动范围
            retractRect.x = Mathf.Clamp(retractRect.x,
                0f, Screen.width - 100f);
            retractRect.y = Mathf.Clamp(retractRect.y,
                0f, Screen.height - 60f);
```

```csharp
        dragableRect = new Rect(0f, 0f, 100f, 20f);
    }
}
private void OnExpandGUI(int windowId)
{
    GUI.DragWindow(dragableRect);
    //关闭窗口
    if (GUILayout.Button("Close", GUILayout.Height(20f)))
        isExpand = false;

    //重绘按钮
    if (GUILayout.Button("Repaint"))
    {
        CollectRoots();
        return;
    }
    scroll = GUILayout.BeginScrollView(scroll);
    for (int i = 0; i < list.Count; i++)
    {
        list[i].Draw();
    }
    GUILayout.EndScrollView();
}
private void OnRetractGUI(int windowId)
{
    GUI.DragWindow(dragableRect);
    //打开窗口
    if (GUILayout.Button("Open", GUILayout.Height(30f)))
        isExpand = true;
}
}
```

4.3.3 运行时检视窗口

当在层级窗口中选择了某个物体后,便可以通过一个检视窗口查看该物体上挂载的组件,通过选择组件,可以查看组件中的详细信息,如图 4-16 所示。

为了实现在这个检视窗口中绘制各类型组件的详细信息,需要定义一个抽象类,该类封装了在检视窗口中绘制文本、开关等控件的方法,是所有组件绘制类的基类,代码如下:

```csharp
//第 4 章/ComponentGUIInspector.cs

using UnityEngine;

public interface IComponentGUIInspector
{
    void Draw(Component component);
```

图 4-16　Inspector GUI Window

```
}

public abstract class ComponentGUIInspector : IComponentGUIInspector
{
    protected string valueStr;
    protected string newValueStr;
    protected float floatValue;
    protected bool boolValue;

    public void Draw(Component component)
    {
        OnDraw(component);
    }
    protected abstract void OnDraw(Component component);

    protected void DrawText(string label, string text, float labelWidth)
    {
        GUILayout.BeginHorizontal();
        GUILayout.Label(label, GUILayout.Width(labelWidth));
        GUILayout.Label(text);
        GUILayout.EndHorizontal();
    }
    protected void DrawText(float spaceWidth, string label,
```

```csharp
        string text, float labelWidth)
{
    GUILayout.BeginHorizontal();
    GUILayout.Space(spaceWidth);
    GUILayout.Label(label, GUILayout.Width(labelWidth));
    GUILayout.Label(text);
    GUILayout.EndHorizontal();
}
protected void DrawText(float spaceWidth, string label, string text)
{
    GUILayout.BeginHorizontal();
    GUILayout.Space(spaceWidth);
    GUILayout.Label(label);
    GUILayout.FlexibleSpace();
    GUILayout.Label(text);
    GUILayout.EndHorizontal();
}

protected bool DrawToggle(string label, bool value, float labelWith)
{
    GUILayout.BeginHorizontal();
    GUILayout.Label(label, GUILayout.Width(labelWith));
    bool retValue = GUILayout.Toggle(value, GUIContent.none);
    GUILayout.EndHorizontal();
    return retValue;
}
protected bool DrawToggle(float spaceWidth, string label,
    bool value, float labelWith)
{
    GUILayout.BeginHorizontal();
    GUILayout.Space(spaceWidth);
    GUILayout.Label(label, GUILayout.Width(labelWith));
    bool retValue = GUILayout.Toggle(value, GUIContent.none);
    GUILayout.EndHorizontal();
    return retValue;
}

protected int DrawInt(string label, int value,
    float labelWith, params GUILayoutOption[] options)
{
    GUILayout.BeginHorizontal();
    GUILayout.Label(label, GUILayout.Width(labelWith));
    valueStr = value.ToString();
    newValueStr = GUILayout.TextField(valueStr, options);
    GUILayout.EndHorizontal();
    if (newValueStr != valueStr)
    {
```

```csharp
            int.TryParse(newValueStr, out value);
        }
        return value;
    }
    protected int DrawInt(float spaceWidth, string label,
        int value, float labelWith, params GUILayoutOption[] options)
    {
        GUILayout.BeginHorizontal();
        GUILayout.Space(spaceWidth);
        GUILayout.Label(label, GUILayout.Width(labelWith));
        valueStr = value.ToString();
        newValueStr = GUILayout.TextField(valueStr, options);
        GUILayout.EndHorizontal();
        if (newValueStr != valueStr)
        {
            int.TryParse(newValueStr, out value);
        }
        return value;
    }

    protected float DrawFloat(string label, float value,
        float labelWith, params GUILayoutOption[] options)
    {
        GUILayout.BeginHorizontal();
        GUILayout.Label(label, GUILayout.Width(labelWith));
        valueStr = value.ToString();
        newValueStr = GUILayout.TextField(valueStr, options);
        GUILayout.EndHorizontal();
        if (newValueStr != valueStr)
        {
            float.TryParse(newValueStr, out value);
        }
        return value;
    }
    protected float DrawFloat(float spaceWidth, string label, float value,
        float labelWith, params GUILayoutOption[] options)
    {
        GUILayout.BeginHorizontal();
        GUILayout.Space(spaceWidth);
        GUILayout.Label(label, GUILayout.Width(labelWith));
        valueStr = value.ToString();
        newValueStr = GUILayout.TextField(valueStr, options);
        GUILayout.EndHorizontal();
        if (newValueStr != valueStr)
        {
            float.TryParse(newValueStr, out value);
        }
```

```csharp
        return value;
    }

    protected Color DrawColor(string label, Color color, float labelWidth)
    {
        GUILayout.BeginHorizontal();
        GUILayout.Label(label, GUILayout.Width(labelWidth));
        GUILayout.Label("R", GUILayout.Width(15f));
        valueStr = (color.r * 255f).ToString();
        newValueStr = GUILayout.TextField(valueStr);
        if (newValueStr != valueStr)
        {
            if (float.TryParse(newValueStr, out floatValue))
            {
                color.r = floatValue / 255f;
            }
        }
        GUILayout.Label("G", GUILayout.Width(15f));
        valueStr = (color.g * 255f).ToString();
        newValueStr = GUILayout.TextField(valueStr);
        if (newValueStr != valueStr)
        {
            if (float.TryParse(newValueStr, out floatValue))
            {
                color.g = floatValue / 255f;
            }
        }
        GUILayout.Label("B", GUILayout.Width(15f));
        valueStr = (color.b * 255f).ToString();
        newValueStr = GUILayout.TextField(valueStr);
        if (newValueStr != valueStr)
        {
            if (float.TryParse(newValueStr, out floatValue))
            {
                color.b = floatValue / 255f;
            }
        }
        GUILayout.Label("A", GUILayout.Width(15f));
        valueStr = (color.a * 255f).ToString();
        newValueStr = GUILayout.TextField(valueStr);
        if (newValueStr != valueStr)
        {
            if (float.TryParse(newValueStr, out floatValue))
            {
                color.a = floatValue / 255f;
            }
        }
```

```csharp
        GUILayout.EndHorizontal();
        return color;
    }

    protected Vector2 DrawVector2(string label, Vector2 vector2,
        float labelWidth, float perFieldWidth)
    {
        GUILayout.BeginHorizontal();
        GUILayout.Label(label, GUILayout.Width(labelWidth));
        GUILayout.Label("X", GUILayout.Width(15f));
        valueStr = vector2.x.ToString();
        newValueStr = GUILayout.TextField(valueStr,
            GUILayout.Width(perFieldWidth));
        if (newValueStr != valueStr)
        {
            if (float.TryParse(newValueStr, out floatValue))
            {
                vector2.x = floatValue;
            }
        }
        GUILayout.Label("Y", GUILayout.Width(15f));
        valueStr = vector2.y.ToString();
        newValueStr = GUILayout.TextField(valueStr,
            GUILayout.Width(perFieldWidth));
        if (newValueStr != valueStr)
        {
            if (float.TryParse(newValueStr, out floatValue))
            {
                vector2.y = floatValue;
            }
        }
        GUILayout.EndHorizontal();
        return vector2;
    }
    protected Vector2 DrawVector2(string label,
        Vector2 vector2, float labelWidth, string x,
        string y, float xyWidth, float perFieldWidth)
    {
        GUILayout.BeginHorizontal();
        GUILayout.Label(label, GUILayout.Width(labelWidth));
        GUILayout.Label(x, GUILayout.Width(xyWidth));
        valueStr = vector2.x.ToString();
        newValueStr = GUILayout.TextField(valueStr,
            GUILayout.Width(perFieldWidth));
        if (newValueStr != valueStr)
        {
            if (float.TryParse(newValueStr, out floatValue))
```

```csharp
        {
            vector2.x = floatValue;
        }
    }
    GUILayout.Label(y, GUILayout.Width(xyWidth));
    valueStr = vector2.y.ToString();
    newValueStr = GUILayout.TextField(valueStr,
        GUILayout.Width(perFieldWidth));
    if (newValueStr != valueStr)
    {
        if (float.TryParse(newValueStr, out floatValue))
        {
            vector2.y = floatValue;
        }
    }
    GUILayout.EndHorizontal();
    return vector2;
}
protected Vector2 DrawVector2(float spaceWidth, string label,
    Vector2 vector2, float labelWidth, float perFieldWidth)
{
    GUILayout.BeginHorizontal();
    GUILayout.Space(spaceWidth);
    GUILayout.Label(label, GUILayout.Width(labelWidth));
    GUILayout.Label("X", GUILayout.Width(15f));
    valueStr = vector2.x.ToString();
    newValueStr = GUILayout.TextField(valueStr,
        GUILayout.Width(perFieldWidth));
    if (newValueStr != valueStr)
    {
        if (float.TryParse(newValueStr, out floatValue))
        {
            vector2.x = floatValue;
        }
    }
    GUILayout.Label("Y", GUILayout.Width(15f));
    valueStr = vector2.y.ToString();
    newValueStr = GUILayout.TextField(valueStr,
        GUILayout.Width(perFieldWidth));
    if (newValueStr != valueStr)
    {
        if (float.TryParse(newValueStr, out floatValue))
        {
            vector2.y = floatValue;
        }
    }
    GUILayout.EndHorizontal();
```

```csharp
        return vector2;
    }
    protected Vector3 DrawVector3(string label, Vector3 vector3,
        float labelWidth, float perFieldWidth)
    {
        GUILayout.BeginHorizontal();
        GUILayout.Label(label, GUILayout.Width(labelWidth));
        GUILayout.Label("X", GUILayout.Width(15f));
        valueStr = vector3.x.ToString();
        newValueStr = GUILayout.TextField(valueStr,
            GUILayout.Width(perFieldWidth));
        if (newValueStr != valueStr)
        {
            if (float.TryParse(newValueStr, out floatValue))
            {
                vector3.x = floatValue;
            }
        }
        GUILayout.Label("Y", GUILayout.Width(15f));
        valueStr = vector3.y.ToString();
        newValueStr = GUILayout.TextField(valueStr,
            GUILayout.Width(perFieldWidth));
        if (newValueStr != valueStr)
        {
            if (float.TryParse(newValueStr, out floatValue))
            {
                vector3.y = floatValue;
            }
        }
        GUILayout.Label("Z", GUILayout.Width(15f));
        valueStr = vector3.z.ToString();
        newValueStr = GUILayout.TextField(valueStr,
            GUILayout.Width(perFieldWidth));
        if (newValueStr != valueStr)
        {
            if (float.TryParse(newValueStr, out floatValue))
            {
                vector3.z = floatValue;
            }
        }
        GUILayout.EndHorizontal();
        return vector3;
    }
    protected float DrawHorizontalSlider(string label,
        float value, float leftValue, float rightValue,
        float labelWidth, float valueLabelWidth)
```

```
    {
        GUILayout.BeginHorizontal();
        GUILayout.Label(label, GUILayout.Width(labelWidth));
        GUILayout.Label(value.ToString("F2"),
            GUILayout.Width(valueLabelWidth));
        value = GUILayout.HorizontalSlider(value, leftValue, rightValue);
        GUILayout.EndHorizontal();
        return value;
    }
    protected float DrawHorizontalSlider(float spaceWidth,
        string label, float value, float leftValue, float rightValue,
        float labelWidth, float valueLabelWidth)
    {
        GUILayout.BeginHorizontal();
        GUILayout.Space(spaceWidth);
        GUILayout.Label(label, GUILayout.Width(labelWidth));
        GUILayout.Label(value.ToString("F2"),
            GUILayout.Width(valueLabelWidth));
        value = GUILayout.HorizontalSlider(value, leftValue, rightValue);
        GUILayout.EndHorizontal();
        return value;
    }
}
```

定义一个特性,用于标记组件绘制类,代码如下:

```
//第4章/ComponentGUIInspectorAttribute.cs

using System;

[AttributeUsage(AttributeTargets.Class)]
public class ComponentGUIInspectorAttribute : Attribute
{
    public Type ComponentType { get; private set; }

    public ComponentGUIInspectorAttribute(Type type)
    {
        ComponentType = type;
    }
}
```

在窗口类中初始化时会根据该特性获取所有类型的组件绘制类,并创建实例将其存储于字典中,当在窗口中选中某个组件时,根据组件类型在字典中获取对应的绘制类实例,如果不存在相应的实例,窗口则会提示"暂不支持该类型组件的调试",代码如下:

```
//第4章/InspectorGUIWindow.cs

using System;
```

```csharp
using System.Linq;
using System.Collections.Generic;

using UnityEngine;

public class InspectorGUIWindow : MonoBehaviour
{
    //层级窗口
    private HierarchyGUIWindow hierarchyGUIWindow;
    private Rect expandRect;
    private Rect retractRect;
    private Rect dragableRect;
    private bool isExpand;
    //当前选中的物体
    private GameObject selected;
    //当前选中物体的组件集合
    private Component[] components;
    private Vector2 listScroll;
    private Vector2 inspectorScroll;
    //当前选中的组件
    private Component currentComponent;
    private Dictionary<string, IComponentGUIInspector> inspectorDic;

    private void Start()
    {
        hierarchyGUIWindow = GetComponent<HierarchyGUIWindow>();
    }

    private void OnEnable()
    {
        inspectorDic = new Dictionary<string, IComponentGUIInspector>();
        var types = GetType().Assembly.GetTypes().Where(
            m => m.IsSubclassOf(typeof(ComponentGUIInspector))).ToArray();
        for (int i = 0; i < types.Length; i++)
        {
            var type = types[i];
            var attributes = type.GetCustomAttributes(false);
            if (attributes.Any(m => m is ComponentGUIInspectorAttribute))
            {
                var target = Array.Find(attributes,
                    m => m is ComponentGUIInspectorAttribute);
                var attribute = target as ComponentGUIInspectorAttribute;
                var instance = Activator.CreateInstance(type);
                inspectorDic.Add(attribute.ComponentType.FullName,
                    instance as IComponentGUIInspector);
            }
        }
    }
```

```csharp
        expandRect = new Rect(0f, 80f, 600f, 500f);
        retractRect = new Rect(0f, 80f, 100f, 60f);
        dragableRect = new Rect(0f, 0f, 600f, 20f);
}
private void OnDisable()
{
    components = null;
    currentComponent = null;
    inspectorDic.Clear();
    inspectorDic = null;
}

private void OnGUI()
{
    if (isExpand)
    {
        expandRect = GUI.Window(2, expandRect,
            OnExpandGUI, "Inspector");
        //限制窗口拖动范围
        expandRect.x = Mathf.Clamp(expandRect.x,
            0f, Screen.width - 600f);
        expandRect.y = Mathf.Clamp(expandRect.y,
            0f, Screen.height - 500f);
        dragableRect = new Rect(0f, 0f, 600f, 20f);
    }
    else
    {
        retractRect = GUI.Window(2, retractRect,
            OnRetractGUI, "Inspector");
        //限制窗口拖动范围
        retractRect.x = Mathf.Clamp(retractRect.x,
            0f, Screen.width - 100f);
        retractRect.y = Mathf.Clamp(retractRect.y,
            0f, Screen.height - 60f);
        dragableRect = new Rect(0f, 0f, 100f, 20f);
    }
}
private void OnExpandGUI(int windowId)
{
    GUI.DragWindow(dragableRect);
    //关闭窗口
    if (GUILayout.Button("Close", GUILayout.Height(20f)))
        isExpand = false;

    if (hierarchyGUIWindow.currentSelected == null)
    {
```

```csharp
            GUILayout.Label("未选中任何物体");
            return;
        }
        if (selected != hierarchyGUIWindow.currentSelected)
        {
            selected = hierarchyGUIWindow.currentSelected;
            components = selected.GetComponents<Component>();
            currentComponent = components[0];
        }
        GUILayout.BeginHorizontal("Box");
        {
            bool active = GUILayout.Toggle(selected.activeSelf,
                string.Empty);
            if (active != selected.activeSelf)
            {
                selected.SetActive(active);
            }
            selected.name = GUILayout.TextField(selected.name,
                GUILayout.Width(Screen.width * .1f));
            GUILayout.FlexibleSpace();
            GUILayout.Label(string.Format("Tag:{0}", selected.tag));
            GUILayout.Space(10f);
            GUILayout.Label(string.Format("Layer:{0}",
                LayerMask.LayerToName(selected.layer)));
        }
        GUILayout.EndHorizontal();

        GUILayout.BeginHorizontal();
        {
            GUILayout.BeginVertical("Box", GUILayout.ExpandHeight(true),
                GUILayout.Width(Screen.width * .075f));
            OnListGUI();
            GUILayout.EndVertical();

            GUILayout.BeginVertical("Box", GUILayout.ExpandHeight(true));
            OnComponentInspector();
            GUILayout.EndVertical();
        }
        GUILayout.EndHorizontal();
    }
    private void OnRetractGUI(int windowId)
    {
        GUI.DragWindow(dragableRect);
        //打开窗口
        if (GUILayout.Button("Open", GUILayout.Height(30f)))
            isExpand = true;
    }
```

```csharp
    private void OnListGUI()
    {
        listScroll = GUILayout.BeginScrollView(listScroll);
        for (int i = 0; i < components.Length; i++)
        {
            if (GUILayout.Toggle(components[i] == currentComponent,
                components[i].GetType().Name))
            {
                currentComponent = components[i];
            }
        }
        GUILayout.EndScrollView();
    }
    private void OnComponentInspector()
    {
        inspectorScroll = GUILayout.BeginScrollView(inspectorScroll);
        string name = currentComponent.GetType().FullName;
        if (inspectorDic.ContainsKey(name))
        {
            inspectorDic[name].Draw(currentComponent);
        }
        else
        {
            GUILayout.Label("暂不支持该类型组件的调试");
        }
        GUILayout.EndScrollView();
    }
}
```

各组件绘制类需要实现抽象类中的抽象方法 OnDraw()，以 Camera 组件的绘制类为例，代码如下：

```csharp
//第4章/GUIInspector.Camera.cs

using UnityEngine;

[ComponentGUIInspector(typeof(Camera))]
public class CameraGUIInspector : ComponentGUIInspector
{
    protected override void OnDraw(Component component)
    {
        Camera camera = component as Camera;

        GUILayout.BeginHorizontal();
        {
            GUILayout.Label("Clear Flags", GUILayout.Width(125f));
```

```csharp
            if (GUILayout.Toggle(camera.clearFlags
                == CameraClearFlags.Skybox, "Skybox"))
                camera.clearFlags = CameraClearFlags.Skybox;
            if (GUILayout.Toggle(camera.clearFlags
                == CameraClearFlags.SolidColor, "SolidColor"))
                camera.clearFlags = CameraClearFlags.SolidColor;
            if (GUILayout.Toggle(camera.clearFlags
                == CameraClearFlags.Depth, "Depth"))
                camera.clearFlags = CameraClearFlags.Depth;
            if (GUILayout.Toggle(camera.clearFlags
                == CameraClearFlags.Nothing, "DontClear"))
                camera.clearFlags = CameraClearFlags.Nothing;
        }
        GUILayout.EndHorizontal();

        camera.backgroundColor = DrawColor("Background",
            camera.backgroundColor, 125f);
        camera.cullingMask = DrawInt("Culling Mask",
            camera.cullingMask, 125f);
        camera.fieldOfView = DrawHorizontalSlider("Field Of View",
            camera.fieldOfView, 1f, 179f, 125f, 40f);
        camera.depth = DrawFloat("Depth", camera.depth, 125f);
    }
}
```

在窗口中选中 Camera 组件，效果如图 4-17 所示。

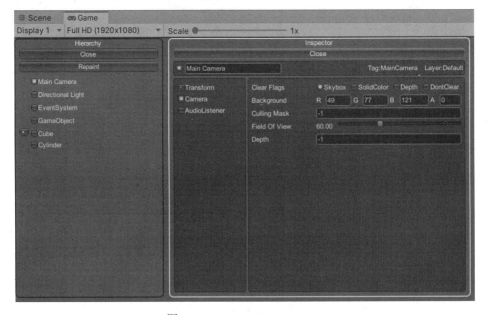

图 4-17　Camera GUI Inspector

第 5 章 编辑器外观

本章将介绍影响编辑器外观的几个重要组成部分，包括 GUI 皮肤、GUI 样式、GUI 图标和 GUI 动画，本章将介绍如何将它们应用于实际的 GUI 设计中。在此之前，先了解 GUI 类中几个与编辑器外观相关的静态变量，见表 5-1。

表 5-1 GUI 类中的静态变量

变量	作用
backgroundColor	用于 GUI 渲染的所有背景元素的着色颜色
color	用于 GUI 渲染的着色颜色
contentColor	用于 GUI 渲染的所有文本的着色颜色
skin	用于 GUI 渲染的 GUISkin 皮肤

5.1 GUI 皮肤

GUISkin 是一种资产文件，选择 Assets→Create→GUI Skin 进行创建，加载该类型资产并通过 GUI.Skin 设置激活对应的皮肤，将 GUI.Skin 设置为 null 表示使用默认的皮肤，如图 5-1 所示。

在该类型资产中存储的是 GUI 各类型控件默认使用的样式和所有自定义的样式，以及相关的设置，见表 5-2。

表 5-2 GUISkin

变量	详解
box	GUI.Box 控件默认使用的样式
button	GUI.Button 控件默认使用的样式
toggle	GUI.Toggle 控件默认使用的样式
label	GUI.Label 控件默认使用的样式
textField	GUI.TextField 控件默认使用的样式
textArea	GUI.TextArea 控件默认使用的样式

续表

变 量	详 解
window	GUI 窗口控件默认使用的样式
horizontalSlider	GUI.HorizontalSlider 控件背景部分默认使用的样式
horizontalSliderThumb	GUI.HorizontalSlider 控件中用于拖动的滑块默认使用的样式
verticalSlider	GUI.VerticalSlider 控件背景部分默认使用的样式
veriticalSliderThumb	GUI.VeriticalSlider 控件中用于拖动的滑块默认使用的样式
horizontalScrollbar	GUI.HorizontalScrollbar 控件背景部分默认使用的样式
horizontalScrollbarThumb	GUI.HorizontalScrollbar 控件中用于拖动的滑块默认使用的样式
horizontalScrollbarLeftButton	GUI.HorizontalScrollbar 控件中的向左按钮默认使用的样式
horizontalScrollbarRightButton	GUI.HorizontalScrollbar 控件中的向右按钮默认使用的样式
veriticalScrollbar	GUI.VerticalScrollbar 控件背景部分默认使用的样式
verticalScrollbarThumb	GUI.VerticalScrollbar 控件中用于拖动的滑块默认使用的样式
verticalScrollbarUpButton	GUI.VerticalScrollbar 控件中的向上按钮默认使用的样式
verticalScrollbarDownButton	GUI.VerticalScrollbar 控件中的向下按钮默认使用的样式
scrollView	GUI 滚动视图控件默认使用的样式
customStyles	所有自定义的样式集合
settings	相关设置

图 5-1 创建 GUISkin 资产

例如，在 Resources 文件夹内创建一个新的 GUISkin 资产，修改与 Label 风格相关的设置，提供一张 Normal 状态下的背景图片，将字体修改为粗体并居中对齐，如图 5-2 所示。

需要注意的是，用于编辑器的相关贴图，需要将其 Texture Type 修改为 Editor GUI and

Legacy GUI,如图 5-3 所示。

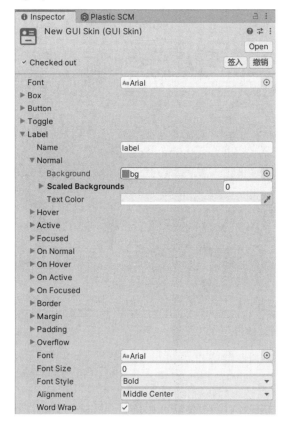

图 5-2 编辑 GUISkin 资产

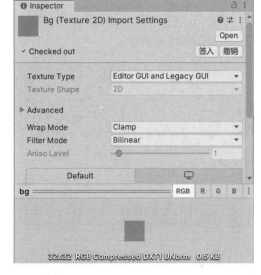

图 5-3 Editor GUI and Legacy GUI

因为该资产位于 Resources 特殊文件夹内,因此可以通过 Resources.Load()方法将该资产加载到内存中,然后获取并使用其中的 Label 风格,代码如下:

```
//第 5 章/GUISkinExampleEditor.cs

using UnityEngine;
using UnityEditor;

[CustomEditor(typeof(GUISkinExample))]
public class GUISkinExampleEditor : Editor
{
    private GUIStyle m_Style;

    public override void OnInspectorGUI()
    {
        base.OnInspectorGUI();
        OnGUIStyleExample();
```

```
    }
    private void OnGUIStyleExample()
    {
        if (m_Style == null)
        {
            GUISkin skin = Resources.Load<GUISkin>("New GUISkin");
            m_Style = skin.label;
        }
        GUILayout.Label("Hello World.", m_Style);
        GUILayout.Label("Today is a good day.", m_Style);
    }
}
```

效果如图 5-4 所示。

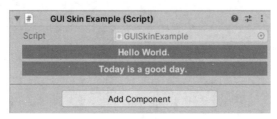

图 5-4　样式效果

5.2　GUI 样式

GUIStyle 是 UnityEngine 命名空间中的类，它用于设置编辑器中控件的样式。在绝大部分绘制控件的方法中提供了包含 GUIStyle 参数的重载方法，参数值可以传入 GUIStyle 类型的变量，也可以通过当前的 GUISkin 中的样式名称设定 GUIStyle，代码如下：

```
GUILayout.Label("Hello World.", new GUIStyle() { alignment = TextAnchor.MiddleLeft, fontStyle = FontStyle.Italic });
GUILayout.Label("Hello World.", "BoldLabel");
```

在第 1 个 Label 文本中，创建了一个新的样式，将其对齐方式设置为居中靠左，将字体样式设置为斜体，除此之外，还可以设置字体大小等。GUISytle 中包含的变量见表 5-3。

表 5-3　GUIStyle 中的变量

变量	详解
alignment	用于设置文本的对齐方式，包含 UpperLeft、UpperCenter、UpperRight、MiddleLeft、MiddleCenter、MiddleRight、LowerLeft、LowerCenter、LowerRight 9 种类型
font	用于设置文本字体
fontSize	用于设置字体的大小
fontStyle	用于设置字体的样式，包含默认、加粗、斜体、加粗斜体共 4 种类型

续表

变量	详解
wordWrap	文本是否应该自动换行
richText	是否为富文本（文本内容是否含标记标签）
textClipping	如何处理要绘制的内容过长而无法放入给定区域的情况，包含 Clip、Overflow 两种类型，分别表示裁剪和溢出
imagePosition	GUIContent 中图像和文本的组合方式，包含 ImageLeft、ImageAbove、ImageOnly、TextOnly 共 4 种类型
contentOffset	要应用该样式的内容的像素偏移
fixedWidth	如果不为 0，则应用该样式的 GUI 元素都将使用该宽度
fixedHeight	如果不为 0，则应用该样式的 GUI 元素都将使用该高度
stretchWidth	是否可以水平拉伸该样式的 GUI 元素来改善布局效果
stretchHeight	是否可以垂直拉伸该样式的 GUI 元素来改善布局效果
normal	正常显示时的渲染设置
hover	鼠标悬浮时的渲染设置
active	按下控件时的渲染设置
focused	聚焦控件时的渲染设置
onNormal	控件处于启用状态时的渲染设置
onHover	控件处于启用状态并且鼠标悬浮在其上方时的渲染设置
onActive	控件处于启用状态并且按下控件时的渲染设置
onFocused	控件处于启用状态并且聚焦控件时的渲染设置
border	背景图片的边框
margin	使用该样式的 GUI 元素与其他 GUI 元素之间的边距
padding	从边缘到内容起始处的空间
overflow	要添加到背景图片的额外空间

在第 2 个 Label 文本中，使用了当前 GUISkin 中的样式名称 BoldLabel 来将其字体样式指定为粗体。通常情况下，在使用默认的 GUISkin 时，开发者并不知道其中包含哪些样式名称，因此需要创建一个工具，以此来查看其中包含的样式。

创建一个新的编辑器窗口，在该窗口中使用滑动列表来列出 GUISkin 中所有的自定义样式，并提供一个输入框，支持检索功能，代码如下：

```
//第 5 章/GUIStylePreviewer.cs

using UnityEngine;
using UnityEditor;

public class GUIStylePreviewer : EditorWindow
{
```

```csharp
[MenuItem("Example/GUIStyle Previewer")]
public static void Open()
{
    GetWindow<GUIStylePreviewer>().Show();
}

private Vector2 scroll;
//检索的内容
private string searchContent = string.Empty;

private void OnGUI()
{
    //搜索栏
    GUILayout.BeginHorizontal(EditorStyles.toolbar);
    GUILayout.Label("Search", GUILayout.Width(50f));
    searchContent = GUILayout.TextField(searchContent,
        EditorStyles.toolbarSearchField);
    GUILayout.EndHorizontal();

    //滚动视图
    scroll = EditorGUILayout.BeginScrollView(scroll);
    for (int i = 0; i < GUI.skin.customStyles.Length; i++)
    {
        var style = GUI.skin.customStyles[i];
        //是否符合检索内容
        if (style.name.ToLower().Contains(searchContent.ToLower()))
        {
            if (GUILayout.Button(style.name, style))
            {
                //当按钮被单击时将样式的名称复制到粘贴板中
                EditorGUIUtility.systemCopyBuffer = style.name;
                Debug.Log(style.name);
            }
        }
    }
    EditorGUILayout.EndScrollView();
}
```

结果如图 5-5 所示。

除了可以使用 GUISkin 中的样式外，EditorStyles 类中也有大量的样式可以使用。同样地，创建一个新的编辑器窗口，在里面列举出 EditorStyles 类中包含的样式，这些样式需要用反射的方式获取，代码如下：

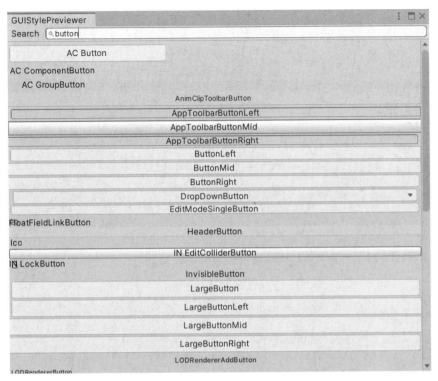

图 5-5 预览默认 GUISkin 中的样式

```
//第 5 章/EditorStylesPreviewer.cs

using System.Linq;
using System.Reflection;

using UnityEngine;
using UnityEditor;

public class EditorStylesPreviewer : EditorWindow
{
    [MenuItem("Example/EditorStyles Previewer")]
    public static void Open()
    {
        GetWindow<EditorStylesPreviewer>().Show();
    }

    private Vector2 scroll;
    private GUIStyle[] styles;

    private void OnEnable()
    {
        //以反射方式获取 EditorStylesPreviewer 中公开的样式
```

```csharp
        PropertyInfo[] propertyInfos = typeof(EditorStyles)
            .GetProperties(BindingFlags.Public | BindingFlags.Static);
        propertyInfos = propertyInfos.Where(m
            => m.PropertyType == typeof(GUIStyle)).ToArray();
        styles = new GUIStyle[propertyInfos.Length];
        for (int i = 0; i < styles.Length; i++)
        {
            styles[i] = propertyInfos[i].GetValue(null, null) as GUIStyle;
        }
    }

    private void OnGUI()
    {
        //滚动视图
        scroll = EditorGUILayout.BeginScrollView(scroll);
        for (int i = 0; i < styles.Length; i++)
        {
            var style = styles[i];
            if (GUILayout.Button(style.name, style))
            {
                //当按钮被单击时将样式的名称复制到粘贴板中
                EditorGUIUtility.systemCopyBuffer = style.name;
                Debug.Log(style.name);
            }
        }
        EditorGUILayout.EndScrollView();
    }
}
```

结果如图 5-6 所示。

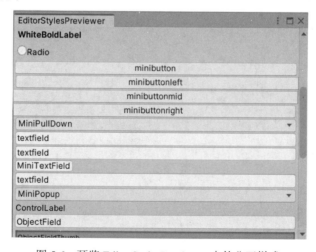

图 5-6 预览 EditorStylesPreviewer 中的公开样式

5.3 GUI 图标

查看 GUILayout 类中的 Label()方法可以发现，其中包含带有 Texture 类型或 GUI Content 类型参数的重载方法，因此 Label 方法不仅可以用于绘制文本内容，还可以用于绘制图片或者图标。例如，加载 Resources 文件夹中名为 Axe 的图片，将它通过 Label()方法绘制到编辑器上，代码如下：

```
//第 5 章/GUISkinExampleEditor.cs

private Texture axe;

private void OnEnable()
{
    axe = Resources.Load<Texture>("Axe");
}
public override void OnInspectorGUI()
{
    base.OnInspectorGUI();
    OnLabelImageExample();
}
private void OnLabelImageExample()
{
    GUILayout.Label(axe, GUILayout.Height(50f));
    GUILayout.Label(new GUIContent("这是一把斧子", axe),
        GUILayout.Height(50f));
}
```

结果如图 5-7 所示。

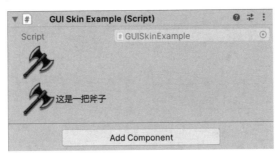

图 5-7　Label 绘制图片

Unity 中内置了大量的图标资产，如果想要使用这些图标，则需要用到 EditorGUIUtility 类中加载内置图标的方法 IconContent()，代码如下，参数 name 表示图标的名称，根据该名称加载 Assets/Editor Default Resources/Icons 中的图标；text 表示鼠标悬浮时的提示文本，需要以"|"字符开头以将其标记为提示文本。

```
public static GUIContent IconContent (string name, string text= null);
```

调用该方法可以加载 Assets/Editor Default Resources/Icons 中的内置图标资产,但是通常情况下,开发者并不知道里面有哪些图标及其对应的名称,它是不可见的,而且在不同的 Unity 版本中里面会有不同的内容,因此创建一个新的编辑器窗口,在编辑器窗口中列举出这些图标。

获取这些图标的名称可以先通过 Resources 中的 FindObjectsOfTypeAll() 方法获取全部贴图资产,然后遍历这些贴图资产,根据其名称调用 IconContent() 方法,如果返回的 GUIContent 中的 image 不为空,则说明资产在内置图标资产的文件夹中。为了防止每次打开编辑器窗口时重新加载,在第 1 次加载时将这些图标名称存储于一个文本文件中并写入缓存,代码如下:

```csharp
//第5章/GUIIconPreviewer.cs
using System.IO;
using System.Text;
using System.Linq;
using System.Collections.Generic;

using UnityEngine;
using UnityEditor;

public sealed class GUIIconPreviewer : EditorWindow
{
    [MenuItem("Example/Built-In GUIIcon Previewer")]
    public static void Open()
    {
        GetWindow<GUIIconPreviewer>().Show();
    }

    private Vector2 scrollPosition;
    private string searchContent = "";
    private const float width = 50f;

    private List<string> iconNames;
    private string[] matchedIconNames;

    private void OnEnable()
    {
        //存储图标名称的文件路径
        string filePath = Path.GetFullPath(".").Replace("\\", "/")
            + "/Library/built-in gui icon names.txt";
        if (!File.Exists(filePath))
        {
            iconNames = new List<string>();
            StringBuilder sb = new StringBuilder();
```

```csharp
            Texture[] textures = Resources.FindObjectsOfTypeAll<Texture>();
            for (int i = 0; i < textures.Length; i++)
            {
                string name = textures[i].name;
                if (string.IsNullOrEmpty(name)) continue;
                if (EditorGUIUtility.IconContent(name).image != null)
                {
                    sb.Append(name);
                    sb.Append("\r");
                    iconNames.Add(name);
                }
            }
            //写入缓存
            File.WriteAllText(filePath, sb.ToString().TrimEnd());
        }
        else
        {
            string fileContent = File.ReadAllText(filePath);
            iconNames = new List<string>(fileContent.Split('\r'));
        }
        matchedIconNames = iconNames.Where(m
            => m.ToLower().Contains(searchContent.ToLower())).ToArray();
    }

    private void OnGUI()
    {
        //搜索栏
        GUILayout.BeginHorizontal(EditorStyles.toolbar);
        {
            GUILayout.Label("Search:", GUILayout.Width(50f));
            EditorGUI.BeginChangeCheck();
            searchContent = GUILayout.TextField(searchContent,
                EditorStyles.toolbarSearchField);
                    //检索的内容发生变更
            if (EditorGUI.EndChangeCheck())
                matchedIconNames = iconNames.Where(m => m.ToLower()
                .Contains(searchContent.ToLower())).ToArray();
        }
        GUILayout.EndHorizontal();

        //滚动视图
        scrollPosition = GUILayout.BeginScrollView(scrollPosition);
        {
            int count = Mathf.RoundToInt(position.width / (width + 3f));
            for (int i = 0; i < matchedIconNames.Length; i += count)
            {
                GUILayout.BeginHorizontal();
```

```
                for (int j = 0; j < count; j++)
                {
                    int index = i + j;
                    if (index < matchedIconNames.Length)
                    {
                        if (GUILayout.Button(EditorGUIUtility
                            .IconContent(matchedIconNames[index]),
                            GUILayout.Width(width), GUILayout.Height(30)))

                        //当按钮被单击时将图标的名称复制到粘贴板中
                        EditorGUIUtility.systemCopyBuffer
                            = matchedIconNames[index];
                        Debug.Log(matchedIconNames[index]);

                    }
                }
                GUILayout.EndHorizontal();
            }
        }
        GUILayout.EndScrollView();
    }
}
```

结果如图 5-8 所示。

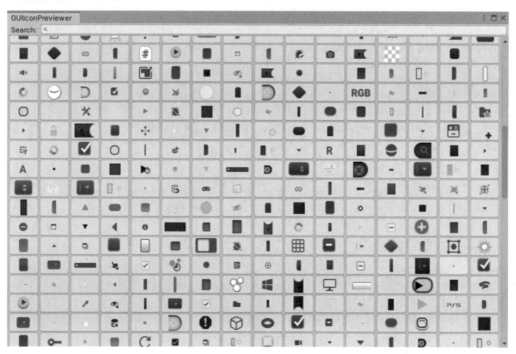

图 5-8 预览内置图标

5.4　GUI 动画

在编辑器界面中可以通过相关的类与方法实现动画效果,以 Canvas 组件为例,当切换 Canvas 组件的 Render Mode 类型时,与该类型相关的序列化属性的交互控件会通过动画的形式呈现出来。

再如 Image 组件,Image Type 由 Simple 切换至 Filled 类型时,Fill Method、Fill Origin、Fill Amount、Clockwise 等序列化属性的交互控件会以从上到下展开的动画效果呈现,如图 5-9 所示。

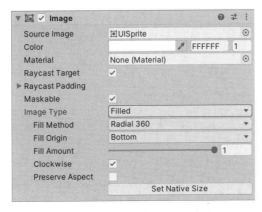

图 5-9　Image 组件的 GUI 动画

动画相关类的抽象基类为 BaseAnimValue,该基类包含的变量见表 5-4。

表 5-4　BaseAnimValue 中的变量

变量	详解
isAnimating	是否正在执行动画过程中
speed	动画的补间速度
target	动画的补间目标
value	动画的当前值
valueChanged	当值发生变更时的回调事件

BaseAnimValue 的派生实现类包括 AnimBool、AnimFloat、AnimQuaternion、AnimVector3。以 AnimBool 为例,可以在其构造函数中设置初始值和回调事件,回调事件也可以通过 valueChanged 添加监听实现,通常会添加 Repaint() 重绘事件。

动画的实现需要使用 EditorGUILayout 类中的 BeginFadeGroup() 方法,它用于开启一个组,以 EndFadeGroup() 方法结束,这个组里的内容通过 BeginFadeGroup() 方法的参数 value 来控制显示或隐藏,取值范围为[0, 1],0 表示隐藏,1 表示完全显示,过渡过程将生成动画,AnimBool 中的 faded 变量将被作为该参数,方法的返回值表示组内内容显示与否,代

码如下:

```csharp
//第5章/GUISkinExampleEditor.cs
private AnimBool animBool;

private void OnEnable()
{
    animBool = new AnimBool(false);
    animBool.valueChanged.AddListener(Repaint);
}
private void OnDisable()
{
    animBool.valueChanged.RemoveListener(Repaint);
}
public override void OnInspectorGUI()
{
    base.OnInspectorGUI();
    OnAnimExample();
}
private void OnAnimExample()
{
    animBool.target = EditorGUILayout.Foldout(animBool.target,
        "这是一个折叠栏", true);
    //当该折叠栏展开或收起时将会有动画效果
    if (EditorGUILayout.BeginFadeGroup(animBool.faded))
    {
        GUILayout.Label("这里是折叠栏里的内容");
        GUILayout.Button("这是一个按钮");
        GUILayout.Toggle(false, "这是一个开关");
    }
    EditorGUILayout.EndFadeGroup();
}
```

展开或收起折叠栏时均会有动画效果，如图 5-10 所示。

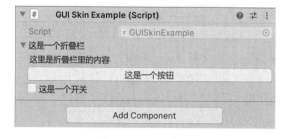

图 5-10　GUI 动画

第 6 章 可视化辅助工具

6.1 Gizmos

6.1.1 概述

在 Scene 窗口和 Game 窗口的右上角均有一个 Gizmos 按钮及下拉列表（不同 Unity 版本可能会有所不同），如图 6-1 所示。Gizmos 按钮开关用于控制窗口中代表组件的 Icon 图标或者被选中组件所绘制的线框的显示或隐藏，可以将其理解为一个可视化辅助工具。

当 3D Icons 开关打开时，组件图标的大小会根据组件所在物体与摄像机的距离进行缩放，使用后面的滑动条控制缩放倍数。图标会被游戏物体遮挡；当开关关闭时，组件图标将会以固定大小进行绘制，后面的滑动条不可交互。图标将会绘制在游戏物体上层，不会被游戏物体遮挡。

Selection Outline 开关用于控制选中游戏物体时，游戏物体中 Mesh 网格的边缘是否高亮。仅可在 Scene 窗口中使用，在 Game 窗口中该选项不可交互。

图 6-1　Gizmos

Selection Wire 开关用于控制选中游戏物体时，游戏物体中 Mesh 网格的线框是否显示。仅可在 Scene 窗口中使用，在 Game 窗口中该选项不可交互。

1. Icon 与 Gizmo

当创建一个新的场景时，场景中会默认包含一个 Main Camera 相机、一个 Directional Light 平行光，以它们为例，当 Gizmos 开启时，它们的 Icon 图标会被绘制在 Scene 窗口中，当选中相机时，相机的视锥体 Gizmo 线框将会被绘制在 Scene 窗口中，如图 6-2 所示。如果不想让指定组件的图标或线框显示，则可以在 Gizmos 下拉列表窗口中找到该组件，取消勾选对应的 Icon 或 Gizmo。

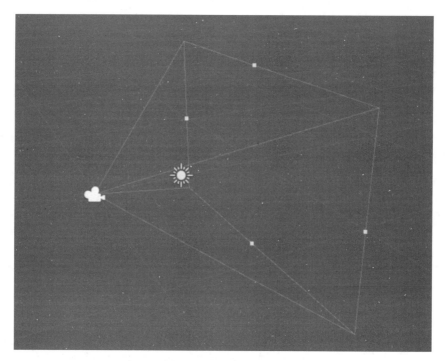

图 6-2 Icon 与 Gizmo

2. Selection Outline 与 Selection Wire

如图 6-3 所示，当选中 Hierarchy 层级中的某个模型物体时，如果 Selection Outline 为开启状态，则可以看到该物体的边缘会出现颜色高亮效果，默认为橙色，也可以在 Edit 菜单中打开 Preferences 偏好设置，找到 Colors 页中的 Selected Outline 进行修改。如果 Selection Wire 为开启状态，则可以看到该物体的 Mesh 网格的线框被绘制出来，线框的颜色同样可以在偏好设置的 Colors 页中进行修改。

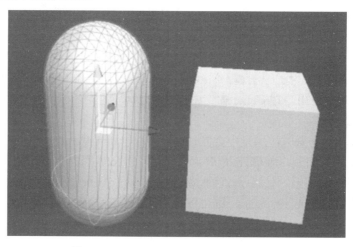

图 6-3 Selection Outline 与 Selection Wire

3. OnDrawGizmos()与 OnDrawGizmosSelected()

OnDrawGizmos()和 OnDrawGizmosSelected()是 MonoBehaviour 中的生命周期函数，它们仅在 Editor 环境中工作，其作用是在场景窗口中绘制辅助图标或线框以实现可视化，两个函数的区别在于，后者仅在对象被选中时才起作用，前者无论对象有没有被选中，始终起作用。绘制主要用到 Gizmos 类中的函数，可以绘制线段，也可以绘制 Cube、Sphere，甚至绘制 Mesh 等，可以绘制实体，也可以绘制线框，示例代码如下：

```
//第6章/GizmosExample.cs

using UnityEngine;

public class GizmosExample : MonoBehaviour
{
    private void OnDrawGizmos()
    {
        Sample();
    }
    private void Sample()
    {
        Gizmos.DrawLine(Vector3.zero, Vector3.up);
        Gizmos.DrawSphere(Vector3.right, .3f);
        Gizmos.DrawWireSphere(Vector3.right * 2f, .3f);
        Gizmos.DrawCube(Vector3.right * 3f, Vector3.one * .3f);
        Gizmos.DrawWireCube(Vector3.right * 4f, Vector3.one * .3f);
    }
}
```

结果如图 6-4 所示。

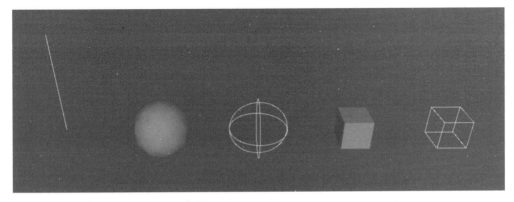

图 6-4　OnDrawGizmos()

4. Draw Gizmo Attribute

MonoBehaviour 中 Gizmos 相关的回调函数用于组件本身的脚本中，而 Draw Gizmo 是一个特性，用于 Editor 脚本中的静态方法，可以为指定类型组件添加 Gizmos 绘制，构造函数的代码如下：

```
public DrawGizmo(GizmoType gizmo, Type drawnGizmoType);
```

参数 gizmo 表示 Gizmo 类型，包含 6 种类型，见表 6-1。参数 drawnGizmoType 表示添加到 Gizmos 绘制的目标组件类型。

表 6-1　GizmoType

类　　型	详　　解
Pickable	是否可以通过选择 Gizmo 去选中相应的 GameObject
NotInSelectionHierarchy	如果本身没有被选中且其父级也没有被选中，则绘制 Gizmo
NonSelected	如果没有被选中，则绘制 Gizmo
Selected	如果已经被选中，则绘制 Gizmo
Active	如果它处于活跃状态（显示在 Inspector 中），则绘制 Gizmo
InSelectionHierarchy	表示如果本身被选中或其父级被选中，则绘制 Gizmo

使用 DrawGizmo 特性的静态方法需要定义两个参数，一个是目标类型组件对象；另一个是 GizmoType 值，示例代码如下：

```
//第 6 章/GizmosExampleEditor.cs

using UnityEngine;
using UnityEditor;

[CustomEditor(typeof(GizmosExample))]
public class GizmosExampleEditor : Editor
{
    [DrawGizmo(GizmoType.Pickable
        | GizmoType.Active | GizmoType.InSelectionHierarchy
        | GizmoType.NotInSelectionHierarchy, typeof(GizmosExample))]
    public static void DrawGizmos(GizmosExample example, GizmoType gizmoType)
    {
        //可以通过 color 变量调整颜色
        Gizmos.color = Color.red;
        Gizmos.DrawLine(Vector3.zero, Vector3.up);
        Gizmos.DrawSphere(Vector3.left, .3f);
        Gizmos.DrawWireSphere(Vector3.left * 2f, .3f);
        Gizmos.DrawCube(Vector3.left * 3f, Vector3.one * .3f);
        Gizmos.DrawWireCube(Vector3.left * 4f, Vector3.one * .3f);
    }
}
```

结果如图 6-5 所示。

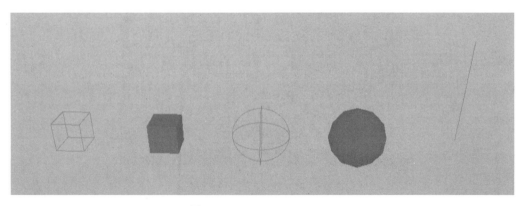

图 6-5　Draw Gizmo Attribute

6.1.2　常用函数

1. DrawLine()与 DrawRay()

这两个函数的作用都是绘制 1 条线段，区别在于参数不同。DrawRay()具有重载方法，转到 DrawRay()方法的实现可以发现其内部实现其实调用的就是 DrawLine()方法。

DrawLine()方法的代码如下，参数 from 表示线段的起点坐标，to 则表示线段的终点坐标。

```
public static void DrawLine(Vector3 from, Vector3 to);
```

DrawRay()方法的代码如下，参数 r 表示射线，from 表示射线的起点坐标，direction 表示射线的方向。

```
public static void DrawRay(Ray r);
public static void DrawRay(Vector3 from, Vector3 direction);
```

示例代码如下：

```
//第 6 章/GizmosExample.cs

private void OnDrawGizmos()
{
    DrawLineRayExample();
}
private void DrawLineRayExample()
{
    Gizmos.DrawLine(Vector3.zero, Vector3.up);
    Gizmos.DrawRay(Vector3.right, Vector3.up);
    Gizmos.DrawRay(new Ray(Vector3.right * 2f, Vector3.up));
}
```

结果如图 6-6 所示。

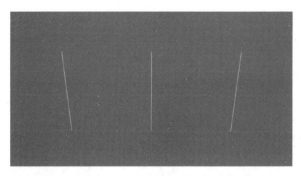

图 6-6　DrawLine()与 DrawRay()

2. DrawCube()与 DrawWireCube()

DrawCube()方法的作用是绘制一个立方体实体，DrawWireCube()方法的作用是绘制一个立方体线框，代码如下，参数 center 表示立方体的中心点坐标，size 表示立方体的大小。

```
public static void DrawCube(Vector3 center, Vector3 size);
public static void DrawWireCube(Vector3 center, Vector3 size);
```

示例代码如下：

```
//第 6 章/GizmosExample.cs

private void OnDrawGizmos()
{
    DrawCubeExample();
}
private void DrawCubeExample()
{
    Gizmos.DrawCube(Vector3.left, Vector3.one * 0.8f);
    Gizmos.DrawWireCube(Vector3.right, Vector3.one * 0.8f);
}
```

结果如图 6-7 所示。

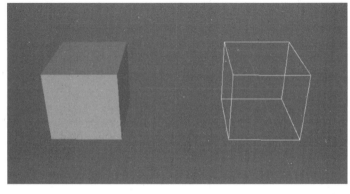

图 6-7　DrawCube()与 DrawWireCube()

3. DrawSphere()与 DrawWireSphere()

DrawSphere()方法的作用是绘制一个球体实体，DrawWireSphere()方法的作用是绘制一个球体线框，代码如下，参数 center 表示球体的中心点坐标，radius 则表示球体的半径。

```
public static void DrawSphere(Vector3 center, float radius);
public static void DrawWireSphere(Vector3 center, float radius);
```

示例代码如下：

```
//第 6 章/GizmosExample.cs

private void OnDrawGizmos()
{
    DrawSphereExample();
}
private void DrawSphereExample()
{
    Gizmos.DrawSphere(Vector3.left, 0.8f);
    Gizmos.DrawWireSphere(Vector3.right, 0.8f);
}
```

结果如图 6-8 所示。

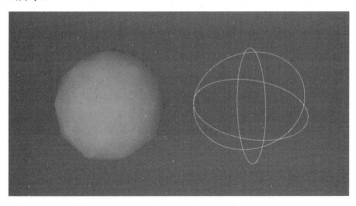

图 6-8　DrawSphere()与 DrawWireSphere()

4. DrawMesh()与 DrawWireMesh()

DrawMesh()方法的作用是绘制一个网格实体，DrawWireMesh()方法的作用是绘制一个网格线框，代码如下，参数 mesh 表示要绘制的网格，position 表示网格的坐标点，rotation 表示网格的旋转朝向，scale 表示网格的缩放值，submeshIndex 表示将绘制的子网格，默认为-1，表示绘制整个网格。

```
public static void DrawMesh (Mesh mesh, Vector3 position, Quaternion rotation,
Vector3 scale);
    public static void DrawMesh (Mesh mesh, int submeshIndex, Vector3 position,
Quaternion rotation, Vector3 scale);
```

```
    public static void DrawWireMesh (Mesh mesh, Vector3 position, Quaternion rotation, Vector3 scale);
    public static void DrawWireMesh (Mesh mesh, int submeshIndex, Vector3 position, Quaternion rotation, Vector3 scale);
```

在 Unity 资源商店中找一个免费的模型资源并导入工程中,如图 6-9 所示,将它放到 Resources 文件夹中。

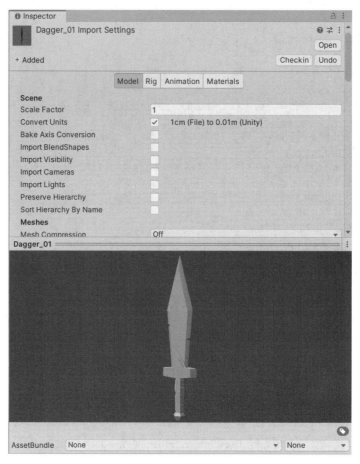

图 6-9　匕首模型

示例代码如下:

```
private Mesh mesh;

private void OnDrawGizmos()
{
    DrawMeshExample();
}
private void DrawMeshExample()
```

```
{
    if (mesh == null)
        mesh = Resources.Load<Mesh>("Dagger_01");
    Gizmos.DrawMesh(mesh, -1, Vector3.left * .3f);
    Gizmos.DrawWireMesh(mesh, -1, Vector3.right * .3f);
}
```

结果如图 6-10 所示。

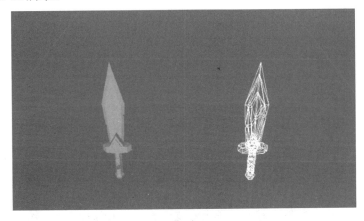

图 6-10 DrawMesh()与 DrawWireMesh()

5. DrawIcon()

DrawIcon()方法的作用是在指定位置绘制一个 Icon 图标，代码如下，参数 center 表示图标绘制的位置；name 表示图标文件的名称，这个图标文件应该存储于 Assets/Gizmos 文件夹中；allowScaling 表示是否允许缩放该图标；tint 表示图标绘制的颜色，默认为白色。

```
public static void DrawIcon(Vector3 center, string name, bool allowScaling);
public static void DrawIcon(Vector3 center, string name, bool allowScaling, Color tint);
```

如图 6-11 所示，在资源商店中找一个斧子类型武器的图标，将它导入工程中并放在 Gizmos 文件夹中。

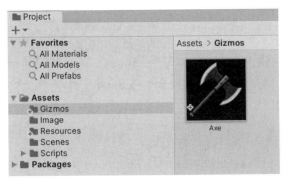

图 6-11 Axe

示例代码如下：

```
//第6章/GizmosExample.cs

private void OnDrawGizmos()
{
    DrawIconExample();
}
private void DrawIconExample()
{
    Gizmos.DrawIcon(Vector3.left, "Axe.png", true);
    Gizmos.DrawIcon(Vector3.right, "Axe.png", true, Color.cyan);
}
```

结果如图 6-12 所示。

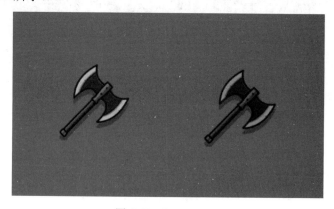

图 6-12　DrawIcon()

6. DrawGUITexture()

DrawGUITexture()方法的作用是绘制一个 Texture 纹理，代码如下，参数 screenRect 表示在 x 轴与 y 轴形成的平面上的纹理大小和位置；texture 表示要绘制的纹理；leftBorder、rightBorder、topBorder、bottomBorder 分别表示左边距、右边距、上边距、下边距；mat 则表示使用此纹理的材质，默认为 null。

```
public static void DrawGUITexture(Rect screenRect, Texture texture);
public static void DrawGUITexture(Rect screenRect, Texture texture, int leftBorder, int rightBorder, int topBorder, int bottomBorder, Material mat);
```

仍然以斧子武器的图片为例，将它放到 Resources 文件夹中，方便使用 Resources.Load()方法进行加载，示例代码如下：

```
//第6章/GizmosExample.cs

private Texture texture;
```

```csharp
private void OnDrawGizmos()
{
    DrawTextureExample();
}
private void DrawTextureExample()
{
    if (texture == null)
        texture = Resources.Load<Texture>("Axe");
    Gizmos.DrawGUITexture(new Rect(0, 0, 1, 1), texture);
}
```

结果如图 6-13 所示。

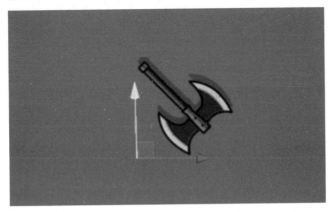

图 6-13　DrawGUITexture()

7. DrawFrustum()

DrawFrustum()方法的作用是绘制一个相机视锥体，代码如下，参数 center 表示绘制的坐标点，即视锥体的顶点；fov 表示视角的大小，以度为单位；maxRange、minRange 分别表示视锥体远平面、近平面的距离；aspect 表示宽高比。

```csharp
public static void DrawFrustum(Vector3 center, float fov, float maxRange,
float minRange, float aspect);
```

示例代码如下：

```csharp
//第 6 章/GizmosExample.cs

private void OnDrawGizmos()
{
    DrawFrustumExample();
}
private void DrawFrustumExample()
{
    Gizmos.DrawFrustum(Vector3.zero, 60f, 10f, 1f, 1.7f);
}
```

结果如图 6-14 所示。

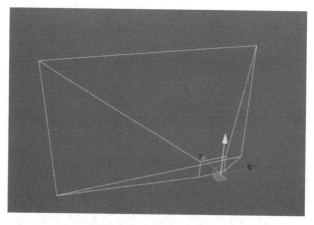

图 6-14　DrawFrustum()

6.1.3　使用 Gizmos 辅助调试相机的避障功能

本节介绍一个相机控制组件，它挂载于 Camera 相机所在的物体，用于实现跟随人物角色，提供第三人称视角控制。

视角的位置受两个参数影响，一个是距离人物角色的高度，另一个是与人物角色的距离，这两个参数首次决定了视角应该所处的位置，为什么说是首次，因为相机与人物角色之间如果有其他的游戏物体，也就是相机看向人物角色时的障碍物，则需要使用相机的避障功能来再次计算相机应该所处的位置，从而避开障碍物，组件代码如下：

```
//第6章/AvatarCameraController.cs

using UnityEngine;

public class AvatarCameraController : MonoBehaviour
{
    //<summary>
    //控制模式
    //</summary>
    public enum ControlMode
    {
        FirstPersonControl,  //第一人称
        ThirdPersonControl,  //第三人称
    }
    //Avatar 角色
    [SerializeField]
    private Transform avatar;
    //控制模式，默认为第三人称
    [SerializeField]
```

```csharp
ControlMode controlMode = ControlMode.ThirdPersonControl;
//切换控制模式的快捷键
[SerializeField]
private KeyCode modeChangeKey;
//视角前方是否与Avatar对齐
[SerializeField]
private bool forwardAlignWithAvatar;
//水平方向灵敏度
[SerializeField, Range(1f, 10f)]
private float horizontalSensitivity = 6f;
//垂直方向灵敏度
[SerializeField, Range(1f, 10f)]
private float verticalSensitivity = 3f;
//用于记录水平方向输入值
private float horizontal;
//用于记录垂直方向输入值
private float vertical;
//旋转x值
private float rotX;
//旋转y值
private float rotY;

//旋转y值的最小值限制
[SerializeField, Range(-80f, -10f)]
private float rotYMinLimit = -40f;
//旋转y值的最大值限制
[SerializeField, Range(10f, 80f)]
private float rotYMaxLimit = 70f;
//插值到目标旋转值所需的时间
[Range(0.01f, 1f), SerializeField]
private float rotationLerpTime = .1f;
//高度
[SerializeField, Range(1f, 5f)]
private float height = 1f;
//默认距离
[SerializeField]
private float distance = 2.5f;
//最小距离限制
[SerializeField, Range(1f, 3f)]
private float minDistanceLimit = 2f;
//最大距离限制
[SerializeField, Range(3f, 10f)]
private float maxDistanceLimit = 5f;
//第一人称模式所用的固定距离
[SerializeField, Range(-1.5f; 0f)]
private float fpmDistance = -.5f;
//鼠标滚轮灵敏度
```

```csharp
[SerializeField, Range(1f, 5f)]
private float scollSensitivity = 2f;
//翻转滚动方向
[SerializeField]
private bool invertScrollDirection;
//与Avatar在水平方向上的偏移值
//仅在forwardAlignWithAvatar为true时开启使用
[SerializeField, Range(-1f, 1f)]
private float horizontalOffset;
//目标距离
private float targetDistance;
//障碍物层级
[SerializeField]
private LayerMask obstacleLayer;
//当为第一人称时是否控制Avatar角色的旋转
[SerializeField]
private bool ctrlAvatarRotWhenFPMode = true;

private void Start()
{
    targetDistance = Mathf.Clamp(distance,
        minDistanceLimit, maxDistanceLimit);
}

private void Update()
{
    if (avatar == null) return;
    if (Input.GetKeyDown(modeChangeKey))
    {
        controlMode = controlMode == ControlMode.FirstPersonControl
            ? ControlMode.ThirdPersonControl
            : ControlMode.FirstPersonControl;
    }
}

private void LateUpdate()
{
    if (avatar == null) return;
    //检测鼠标右键按下
    if (Input.GetMouseButton(1))
    {
        horizontal = forwardAlignWithAvatar ? 0f : Input.GetAxis("Mouse X")
            * Time.deltaTime* 100f * horizontalSensitivity;
        vertical = Input.GetAxis("Mouse Y")
            * Time.deltaTime * 100f * verticalSensitivity;

        rotX += horizontal;
```

```csharp
        rotY -= vertical;
        //限制旋转 y 值角度
        rotY = Mathf.Clamp(rotY, rotYMinLimit, rotYMaxLimit);
    }

    //通过鼠标滚轮的滚动改变距离
    distance -= Input.GetAxis("Mouse ScrollWheel") * Time.deltaTime
        * 100f * scollSensitivity * (invertScrollDirection ? -1f : 1f);
    //距离限制
    distance = Mathf.Clamp(distance, minDistanceLimit,
        maxDistanceLimit);
    //以插值方式计算距离
    targetDistance = controlMode == ControlMode.ThirdPersonControl
        ? Mathf.Lerp(targetDistance, distance, Time.deltaTime
            * scollSensitivity) : fpmDistance;

    //目标旋转值
    Quaternion targetRotation = Quaternion.Euler(rotY, rotX, 0f);
    //旋转值插值率
    float rotationLerpPct = 1f - Mathf.Exp(Mathf.Log(1f - .99f)
        / rotationLerpTime * Time.deltaTime);
    //以插值方式计算旋转值
    targetRotation = Quaternion.Lerp(transform.rotation,
        targetRotation, rotationLerpPct);

    //目标坐标值
    Vector3 targetPosition = targetRotation * Vector3.forward
        * -targetDistance + avatar.position + Vector3.up * height;
    //避障
    targetPosition = ObstacleAvoidance(targetPosition,
        avatar.position + Vector3.up * height, .1f, distance);

    transform.position = targetPosition + Vector3.left
        * horizontalOffset;
    transform.rotation = targetRotation;

    //第一人称控制模式，相机视角旋转的同时控制 Avatar 角色的旋转
    if (controlMode == ControlMode.FirstPersonControl
        && ctrlAvatarRotWhenFPMode)
    {
        Vector3 euler = Vector3.zero;
        //只取相机的 RotY
        euler.y = targetRotation.eulerAngles.y;
        avatar.rotation = Quaternion.Euler(euler);
    }
}
```

```csharp
//避障
private Vector3 ObstacleAvoidance(Vector3 current, Vector3 target,
    float radius, float maxDistance)
{
    Ray ray = new Ray(target, current - target);
    if (Physics.SphereCast(ray, radius, out RaycastHit hit,
        maxDistance, obstacleLayer))
    {
        return ray.GetPoint(hit.distance - radius * 2f);
    }
    return current;
}
```

为该组件实现 OnDrawGizmos()方法，在相机的避障功能生效时，使用 Gizmos 中的 DrawWireSphere()方法绘制出调用避障函数之前相机应该所在的位置，然后绘制出避障之后相机应该所在的位置，以此来辅助调试该组件的避障功能，代码如下：

```csharp
//第 6 章/AvatarCameraController.cs
private void OnDrawGizmos()
{
    Quaternion targetRotation = Quaternion.Euler(rotY, rotX, 0f);
    float rotationLerpPct = 1f - Mathf.Exp(Mathf.Log(1f - .99f)
        / rotationLerpTime * Time.deltaTime);
    targetRotation = Quaternion.Lerp(transform.rotation,
        targetRotation, rotationLerpPct);
    Vector3 targetPosition = targetRotation * Vector3.forward
        * -targetDistance + avatar.position + Vector3.up * height;
    Vector3 target = avatar.position + Vector3.up * height;
    Ray ray = new Ray(target, targetPosition - target);
    if (Physics.SphereCast(ray, .1f, out RaycastHit hit,
        distance, obstacleLayer))
    {
        Gizmos.color = Color.red;
        Gizmos.DrawWireSphere(targetPosition, .1f);
        Gizmos.color = Color.green;
        Vector3 pos = ray.GetPoint(hit.distance - .2f);
        Gizmos.DrawWireSphere(pos, .1f);
        Gizmos.color = Color.white;
        Gizmos.DrawLine(targetPosition, pos);
    }
}
```

结果如图 6-15 所示。

图 6-15　使用 Gizmos 辅助调试相机的避障功能

在这个功能中使用了 Shere Cast 物理检测，而 Physics 中其他类型的物理检测，例如 Ray Cast、Box Cast 在开发中也会被大量使用，因此也可以使用 Gizmos 中相应的方法（如 DrawLine()、DrawWireCube()）实现物理检测的可视化调试。

6.2　Handles

6.2.1　概述

Handles 类的作用类似于 Gizmos，同样是可视化辅助工具类，可以在场景视图中绘制点、线、面等几何图形，不同的是，除此之外，Handles 还可以提供用户交互的操控柄，常见的操控柄有坐标操控柄、旋转操控柄、缩放操控柄等。

该类主要用于自定义 Editor 编辑器类中的 OnSceneGUI()回调方法，如果想在编辑器窗口中使用，则可以使用 SceneView 类中的 duringSceneGui 进行事件注册，示例代码如下：

```
//第6章/SceneGUIExample.cs

using UnityEngine;
using UnityEditor;

public class SceneGUIExample : EditorWindow
{
    [MenuItem("Example/Scene GUI Example")]
    public static void Open()
```

```
        {
            GetWindow<SceneGUIExample>().Show();
        }

    private void OnEnable()
    {
        SceneView.duringSceneGui += OnSceneGUI;
    }
    private void OnDisable()
    {
        SceneView.duringSceneGui -= OnSceneGUI;
    }
    private void OnSceneGUI(SceneView sceneView)
    {
        Handles.DrawLine(Vector3.zero, Vector3.up);
        //TOOD
    }
}
```

6.2.2 常用函数

1. DrawLine()

DrawLine()方法的作用是根据两个坐标点绘制 1 条线段,代码如下,参数 p1 表示线段的起点,p2 表示线段的终点,thickness 表示线段的粗细,默认值为 0,表示单像素。

```
public static void DrawLine(Vector3 p1, Vector3 p2, float thickness);
```

示例代码如下:

```
//第 6 章/HandlesExampleEditor.cs

using UnityEngine;
using UnityEditor;

[CustomEditor(typeof(HandlesExample))]
public class HandlesExampleEditor : Editor
{
    private void OnSceneGUI()
    {
        DrawLineExample();
    }
    private void DrawLineExample()
    {
        Handles.color = Color.cyan;
        for (int i = 0; i < 10; i++)
        {
            Vector3 p1 = Vector3.right * (i * 0.5f);
```

```
            Vector3 p2 = p1 + Vector3.up;
            Handles.DrawLine(p1, p2, i);
        }
    }
}
```

结果如图 6-16 所示。

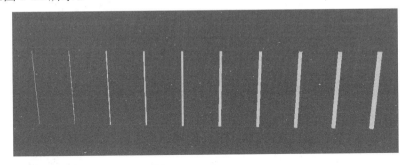

图 6-16　DrawLine()

2. DrawPolyLine()

DrawPolyLine()方法的作用是根据一系列坐标点绘制线段，前后坐标点之间两两相连，代码如下：

```
public static void DrawPolyLine(params Vector3[] points);
```

在 HandlesExample 组件中声明一个 Transform 类型数组 points，在场景中创建一些 cube，将它们的 Transform 放入 points 中，在编辑器中根据这些 Transform 的坐标位置绘制线段，代码如下：

```
//第 6 章/HandlesExampleEditor.cs

private HandlesExample example;

private void OnEnable()
{
    example = target as HandlesExample;
}
private void OnSceneGUI()
{
    DrawPolyLineExample();
}
private void DrawPolyLineExample()
{
    Vector3[] points = new Vector3[example.points.Length];
    for (int i = 0; i < points.Length; i++)
    {
        points[i] = example.points[i].position;
```

```
        }
        Handles.DrawPolyLine(points);
}
```

结果如图 6-17 所示。

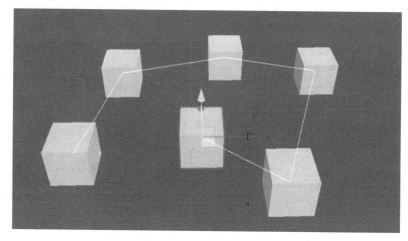

图 6-17　DrawPolyLine()

3. DrawWireCube()

与 Gizmos 中 DrawWireCube() 方法的作用与用法一致,用于绘制一个立方体形状的线框,代码如下,参数 center 表示立方体的中心点坐标,size 表示立方体的大小。

```
public static void DrawWireCube(Vector3 center, Vector3 size);
```

示例代码如下:

```
//第 6 章/HandlesExampleEditor.cs

private void OnSceneGUI()
{
    DrawWireCubeExample();
}
private void DrawWireCubeExample()
{
    Handles.color = Color.cyan;
    Handles.DrawWireCube(Vector3.zero, Vector3.one);
}
```

结果如图 6-18 所示。

4. DrawWireArc() 与 DrawSolidArc()

DrawWireArc() 方法的作用是绘制一条圆弧线,而 DrawSolidArc() 方法的作用是绘制一个实心的圆弧,也就是一个扇形,代码如下,参数 center 表示圆的中心点坐标;normal 表示圆的法线方向;from 表示圆周上的点相对于圆心的方向,即起点;angle 表示圆弧的角度;

图 6-18　DrawWireCube()

radius 表示圆的半径；thickness 表示圆弧线的粗细，默认值为 0，表示单像素。

```
public static void DrawWireArc (Vector3 center, Vector3 normal, Vector3 from,
float angle, float radius, float thickness);
public static void DrawSolidArc (Vector3 center, Vector3 normal, Vector3 from,
float angle, float radius);
```

示例代码如下：

```
//第 6 章/HandlesExampleEditor.cs

private void OnSceneGUI()
{
    DrawArcExample();
}
private void DrawArcExample()
{
    Handles.DrawWireArc(Vector3.left, Vector3.up,
        Vector3.left, 180f, 1f);
    Handles.DrawSolidArc(Vector3.right, Vector3.up,
        Vector3.left, 180f, 1f);
}
```

结果如图 6-19 所示。

5. DrawWireDisc()与 DrawSolidDisc()

DrawWireDisc()方法的作用是绘制一个圆形线框，相应地，DrawSolidDisc()方法的作用是绘制一个实心的圆盘，代码如下，参数 center 表示圆的中心点坐标；normal 表示圆的法线方向；radius 表示圆的半径；thickness 表示圆形线框的粗细，默认值为 0，表示单像素。

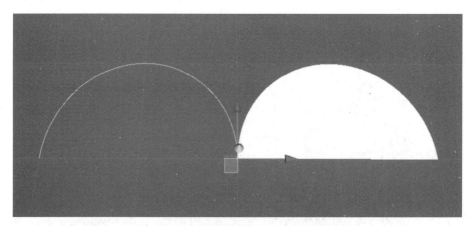

图 6-19　DrawWireArc()与 DrawSolidArc()

```
    public static void DrawWireDisc (Vector3 center, Vector3 normal, float radius,
float thickness);
    public static void DrawSolidDisc(Vector3 center, Vector3 normal, float
radius);
```

示例代码如下：

```
//第6章/HandlesExampleEditor.cs

private void OnSceneGUI()
{
    DrawDiscExample();
}
private void DrawDiscExample()
{
    Handles.DrawWireDisc(Vector3.left, Vector3.up, 1f);
    Handles.DrawSolidDisc(Vector3.right, Vector3.up, 1f);
}
```

结果如图 6-20 所示。

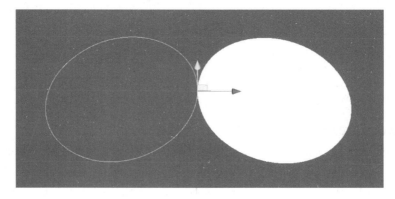

图 6-20　DrawWireDisc()与 DrawSolidDisc()

6. PositionHandle()

PositionHandle()方法用于在场景视图创建一个坐标操控柄,方法返回值是用户与操控柄交互修改后的新值,代码如下,参数 position 表示操控柄的坐标,rotation 表示操控柄的方向。

```
public static Vector3 PositionHandle (Vector3 position, Quaternion rotation);
```

例如在 HandlesExample 组件中声明一个 Vector3 类型的变量,在其编辑器类 HandlesExampleEditor 中的 OnSceneGUI()方法中通过 PositionHandle()方法来控制这个变量的值,代码如下:

```
//第 6 章/HandlesExampleEditor.cs
private void OnSceneGUI()
{
    PositionHandleExample();
}
private void PositionHandleExample()
{
    Handles.Label(example.pos, "拖动这个操控柄修改 pos 的值");
    example.pos = Handles.PositionHandle(example.pos, Quaternion.identity);
}
```

结果如图 6-21 所示,左侧的操控柄是选中 HandlesExample 组件时它本身的坐标操控柄,而右侧的操控柄是在 OnSceneGUI()方法中创建的用于控制 pos 变量的坐标操控柄,拖动它可以改变 pos 变量的值。

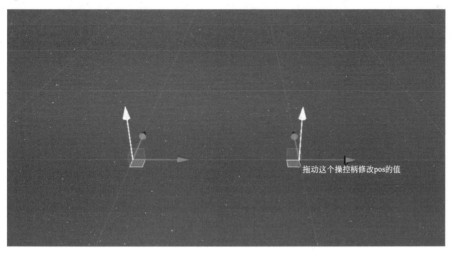

图 6-21　PositionHandle()

7. RotationHandle()

RotationHandle()方法用于在场景视图创建一个旋转操控柄,此方法的返回值是用户与操控柄交互修改后的新值，代码如下，参数 rotation 表示要修改的方向值，position 表示操控柄的坐标。

```
public static Quaternion RotationHandle (Quaternion rotation, Vector3 position);
```

在 HandlesExample 组件中再声明一个表示旋转值的 Vector3 类型的变量，在编辑器类中通过 RotationHandle() 方法来控制这个变量的值，代码如下：

```
//第6章/HandlesExampleEditor.cs

private void OnSceneGUI()
{
    RotationHandleExample();
}
private void RotationHandleExample()
{
    Handles.Label(Vector3.right,
        string.Format("Rot: {0}", example.rot));
    example.rot = Handles.RotationHandle(
        Quaternion.Euler(example.rot), Vector3.right).eulerAngles;
}
```

结果如图 6-22 所示。

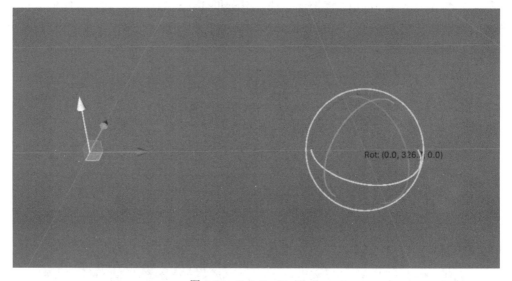

图 6-22　RotationHandle()

8. ScaleHandle()

ScaleHandle() 方法用于在场景视图创建一个缩放操控柄，此方法的返回值是用户与操控柄交互修改后的新值，代码如下，参数 scale 表示要修改的缩放值，position 表示操控柄的坐标，rotation 表示操控柄的方向，size 表示操控柄的大小，如果希望操控柄在屏幕上保持固定的大小，则可以通过 HandleUtility 类中的 GetHandleSize() 方法计算大小。

```
public static Vector3 ScaleHandle (Vector3 scale, Vector3 position,
Quaternion rotation, float size);
```

示例代码如下：

```
//第6章/HandlesExampleEditor.cs

private void OnSceneGUI()
{
    ScaleHandleExample();
}
private void ScaleHandleExample()
{
    Handles.Label(Vector3.right,
        string.Format("Scale: {0}", example.scale));
    example.scale = Handles.ScaleHandle(
        example.scale, Vector3.right, Quaternion.identity,
        HandleUtility.GetHandleSize(Vector3.right));
}
```

结果如图 6-23 所示。

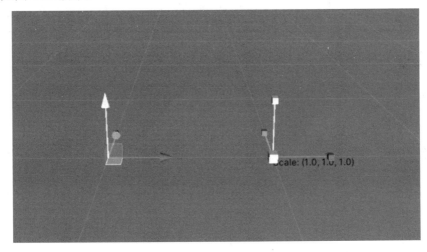

图 6-23　ScaleHandle()

9. RadiusHandle()

RadiusHandle() 方法用于在场景视图中创建一个半径操控柄，此方法的返回值是用户与操控柄交互修改后的新值，代码如下，参数 rotation 表示操控柄的方向；position 表示操控柄的坐标；radius 表示要修改的半径；handlesOnly 表示是否忽略半径的圆形轮廓并且只绘制点手柄，默认值为 false。

```
public static float RadiusHandle(Quaternion rotation, Vector3 position, float
radius, bool handlesOnly);
```

示例代码如下:

```
//第 6 章/HandlesExampleEditor.cs

private void OnSceneGUI()
{
    RadiusHandleExample();
}
private void RadiusHandleExample()
{
    example.radius = Handles.RadiusHandle(
        Quaternion.identity, Vector3.left, example.radius);
    example.radius = Handles.RadiusHandle(
        Quaternion.identity, Vector3.right, example.radius, true);
}
```

结果如图 6-24 所示。

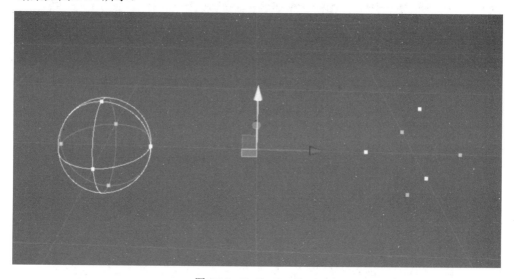

图 6-24　RadiusHandle()

10. Label()

Label()方法用于在场景视图中创建一个文本标签,此方法具有多个重载,代码如下,参数 position 表示标签的坐标; text 表示要在标签上显示的文本; image 表示要在标签上显示的纹理; content 表示要在标签上显示的文本、图像和提示; style 表示要使用的样式,默认为当前 GUISkin 中的 Label 样式。

```
public static void Label (Vector3 position, string text);
public static void Label (Vector3 position, Texture image);
```

```
public static void Label (Vector3 position, GUIContent content);
public static void Label (Vector3 position, string text, GUIStyle style);
public static void Label (Vector3 position, GUIContent content, GUIStyle style);
```

示例代码如下:

```
//第 6 章/HandlesExampleEditor.cs
private void OnSceneGUI()
{
    LabelExample();
}
private void LabelExample()
{
    Handles.Label(Vector3.left * .5f, "Hello World.");
    Handles.Label(Vector3.right * .5f, "Hello World.",
        EditorStyles.whiteLabel);
}
```

结果如图 6-25 所示。

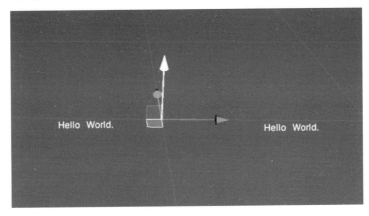

图 6-25　Label()

11. HandleCap()

Handles 类中包含多种类型的 HandleCap()绘制方法,这些方法的参数一致,因此以 ArrowHandleCap()为例,代码如下,参数 controlID 表示操控柄的控件 ID;position 表示操控柄的坐标;rotation 表示操控柄的方向;size 表示操控柄的大小;eventType 表示事件类型。

```
public static void ArrowHandleCap (int controlID, Vector3 position,
Quaternion rotation, float size, EventType eventType);
```

示例代码如下:

```
//第 6 章/HandlesExampleEditor.cs
```

```csharp
private void OnSceneGUI()
{
    HandleCapExample();
}
private void HandleCapExample()
{
    if (Event.current.type == EventType.Repaint)
    {
        Transform transform = example.transform;
        float size = HandleUtility.GetHandleSize(transform.position);
        Handles.color = Handles.xAxisColor;
        Handles.CircleHandleCap(0, transform.position
            + new Vector3(3f, 0f, 0f), transform.rotation
            * Quaternion.LookRotation(Vector3.right), size,
            EventType.Repaint);
        Handles.ArrowHandleCap(0, transform.position
            + new Vector3(3f, 0f, 0f), transform.rotation
            * Quaternion.LookRotation(Vector3.right), size,
            EventType.Repaint);
        Handles.RectangleHandleCap(0, transform.position
            + new Vector3(6f, 0f, 0f), transform.rotation
            * Quaternion.LookRotation(Vector3.right), size,
            EventType.Repaint);
        Handles.SphereHandleCap(0, transform.position
            + new Vector3(8f, 0f, 0f), transform.rotation
            * Quaternion.LookRotation(Vector3.right), size,
            EventType.Repaint);
        Handles.color = Handles.yAxisColor;
        Handles.CircleHandleCap(0, transform.position
            + new Vector3(0f, 3f, 0f), transform.rotation
            * Quaternion.LookRotation(Vector3.up), size,
            EventType.Repaint);
        Handles.ArrowHandleCap(0, transform.position
            + new Vector3(0f, 3f, 0f), transform.rotation
            * Quaternion.LookRotation(Vector3.up), size,
            EventType.Repaint);
        Handles.RectangleHandleCap(0, transform.position
            + new Vector3(0f, 6f, 0f), transform.rotation
            * Quaternion.LookRotation(Vector3.up), size,
            EventType.Repaint);
        Handles.SphereHandleCap(0, transform.position
            + new Vector3(0f, 8f, 0f), transform.rotation
            * Quaternion.LookRotation(Vector3.up), size,
            EventType.Repaint);
        Handles.color = Handles.zAxisColor;
        Handles.CircleHandleCap(0, transform.position
            + new Vector3(0f, 0f, 3f), transform.rotation
```

```
            * Quaternion.LookRotation(Vector3.forward), size,
            EventType.Repaint);
        Handles.ArrowHandleCap(0, transform.position
            + new Vector3(0f, 0f, 3f), transform.rotation
            * Quaternion.LookRotation(Vector3.forward), size,
            EventType.Repaint);
        Handles.RectangleHandleCap(0, transform.position
            + new Vector3(0f, 0f, 6f), transform.rotation
            * Quaternion.LookRotation(Vector3.forward), size,
            EventType.Repaint);
        Handles.SphereHandleCap(0, transform.position
            + new Vector3(0f, 0f, 8f), transform.rotation
            * Quaternion.LookRotation(Vector3.forward), size,
            EventType.Repaint);
    }
}
```

结果如图 6-26 所示。

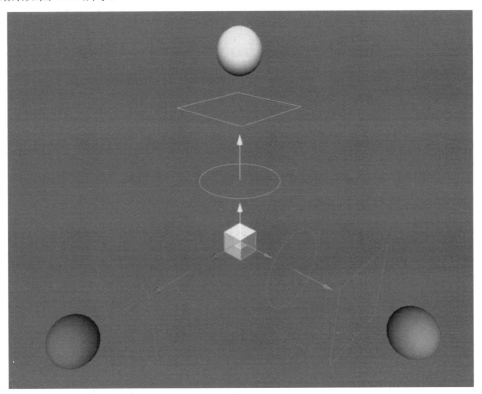

图 6-26 HandleCap()

6.2.3 实现一个路径编辑工具

本节以一个路径编辑工具为例具体展示 Handles 类的使用，工具的实现主要用到了贝塞尔曲线（Bezier Curve），贝塞尔曲线是计算机图形学中相当重要的参数曲线。

在 Unity 中通过一系列的坐标点和与坐标点形成切线的控制点生成一条贝塞尔曲线，而这些坐标点或控制点通过 Handles 类中的 PositionHandle 坐标操控柄来控制，最终实现路径的编辑。

1. 贝塞尔曲线公式

1）一阶贝塞尔曲线

给定点 P_0、P_1，一阶贝塞尔曲线只是一条两点之间的直线，公式及实现代码如下：

$$B(t) = P_0 + (P_1 - P_0)\, t = (1-t)\, P_0 + tP_1, \quad t \in [0, 1] \tag{6-1}$$

```csharp
//第 6 章/BezierCurveUtility.cs

//<summary>
//一阶贝塞尔曲线
//</summary>
//<param name="p0">起点</param>
//<param name="p1">终点</param>
//<param name="t">[0,1]</param>
//<returns></returns>
public static Vector3 Bezier1(Vector3 p0, Vector3 p1, float t)
{
    return (1 - t) * p0 + t * p1;
}
```

使用 Handles 和 Gizmos 类在 OnDrawGizmos 中绘制一阶贝塞尔曲线，代码如下：

```csharp
//第 6 章/BezierCurvePathExample.cs

using UnityEngine;
using UnityEditor;

public class BezierCurvePathExample : MonoBehaviour
{
    private float t;

    private void Update()
    {
        if (t < 1f)
        {
            t += Time.deltaTime * .2f;
            t = Mathf.Clamp01(t);
        }
    }
```

```csharp
#if UNITY_EDITOR
    private void OnDrawGizmos()
    {
        Bezier1Example();
    }
    private void Bezier1Example()
    {
        Gizmos.color = Color.grey;
        Vector3 p0 = Vector3.left * 5f;
        Vector3 p1 = Vector3.right * 5f;
        Gizmos.DrawLine(p0, p1);
        Handles.Label(p0, "P0");
        Handles.Label(p1, "P1");
        Handles.SphereHandleCap(0, p0, Quaternion.identity,
            0.1f, EventType.Repaint);
        Handles.SphereHandleCap(0, p1, Quaternion.identity,
            0.1f, EventType.Repaint);
        Vector3 pt = BezierCurveUtility.Bezier1(p0, p1, t);
        Gizmos.color = Color.red;
        Gizmos.DrawLine(p0, pt);
        Handles.Label(pt, string.Format("Pt (t = {0})", t));
        Handles.SphereHandleCap(0, pt, Quaternion.identity,
            0.1f, EventType.Repaint);
    }
#endif
}
```

运行程序，结果如图 6-27 所示。

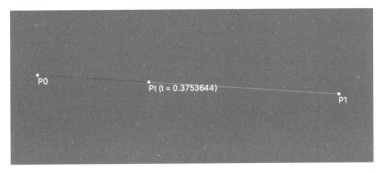

图 6-27　一阶贝塞尔曲线

2）二阶贝塞尔曲线

路径由给定点 P_0、P_1、P_2 的函数计算，公式及实现代码如下：

$$B(t) = (1-t)^2 P_0 + 2t(1-t) P_1 + t^2 P_2, \quad t \in [0, 1] \tag{6-2}$$

```csharp
//第 6 章/BezierCurveUtility.cs

//<summary>
//二阶贝塞尔曲线
//</summary>
//<param name="p0">起点</param>
//<param name="p1">控制点</param>
//<param name="p2">终点</param>
//<param name="t">[0,1]</param>
//<returns></returns>
public static Vector3 Bezier2(Vector3 p0, Vector3 p1, Vector3 p2, float t)
{
    Vector3 p0p1 = (1 - t) * p0 + t * p1;
    Vector3 p1p2 = (1 - t) * p1 + t * p2;
    return (1 - t) * p0p1 + t * p1p2;
}
```

使用 Handles 和 Gizmos 类在 OnDrawGizmos 中绘制二阶贝塞尔曲线，代码如下：

```csharp
//第 6 章/BezierCurvePathExample.cs

private void OnDrawGizmos()
{
    Bezier2Example();
}
private void Bezier2Example()
{
    Gizmos.color = Color.grey;
    Vector3 p0 = Vector3.left * 5f;
    Vector3 p1 = Vector3.left * 2f + Vector3.forward * 2f;
    Vector3 p2 = Vector3.right * 5f;
    Gizmos.DrawLine(p0, p1);
    Gizmos.DrawLine(p2, p1);
    Handles.Label(p0, "P0");
    Handles.Label(p1, "P1");
    Handles.Label(p2, "P2");
    Handles.SphereHandleCap(0, p0, Quaternion.identity,
        0.1f, EventType.Repaint);
    Handles.SphereHandleCap(0, p1, Quaternion.identity,
        0.1f, EventType.Repaint);
    Handles.SphereHandleCap(0, p2, Quaternion.identity,
        0.1f, EventType.Repaint);

    Gizmos.color = Color.green;
    for (int i = 0; i < 100; i++)
    {
        Vector3 curr = BezierCurveUtility.Bezier2(
```

```
            p0, p1, p2, i / 100f);
        Vector3 next = BezierCurveUtility.Bezier2(
            p0, p1, p2, (i + 1) / 100f);
        Gizmos.color = t > (i / 100f)
            ? Color.red : Color.green;
        Gizmos.DrawLine(curr, next);
    }
    Vector3 pt = BezierCurveUtility.Bezier2(p0, p1, p2, t);
    Handles.Label(pt, string.Format("Pt (t = {0})", t));
    Handles.SphereHandleCap(0, pt, Quaternion.identity,
        0.1f, EventType.Repaint);
}
```

运行程序，结果如图6-28所示。

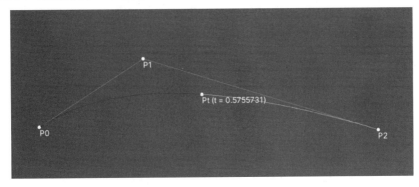

图 6-28 二阶贝塞尔曲线

3) 三阶贝塞尔曲线

P_0、P_1、P_2、P_3 这 4 个点在平面或三维空间中定义了 3 次方贝塞尔曲线。曲线起始于 P_0 走向 P_1，并从 P_2 的方向来到 P_3，一般不会经过 P_1、P_2，这两个点只提供方向信息，可以将 P_1、P_2 理解为控制点。P_0 和 P_1 之间的间距决定了曲线在转而趋近 P_3 之前走向 P_2 的长度有多长，公式及实现代码如下：

$$B(t) = P_0(1-t)^3 + 3P_1 t(1-t)^2 + 3P_2 t^2(1-t) + P_3 t^3, \quad t \in [0, 1] \tag{6-3}$$

```
//第 6 章/BezierCurveUtility.cs

//<summary>
//三阶贝塞尔曲线
//</summary>
//<param name="p0">起点</param>
//<param name="p1">控制点 1</param>
//<param name="p2">控制点 2</param>
//<param name="p3">终点</param>
//<param name="t">[0,1]</param>
//<returns></returns>
```

```csharp
public static Vector3 Bezier3(Vector3 p0, Vector3 p1, Vector3 p2, Vector3 p3, float t)
{
    Vector3 p0p1 = (1 - t) * p0 + t * p1;
    Vector3 p1p2 = (1 - t) * p1 + t * p2;
    Vector3 p2p3 = (1 - t) * p2 + t * p3;
    Vector3 p0p1p2 = (1 - t) * p0p1 + t * p1p2;
    Vector3 p1p2p3 = (1 - t) * p1p2 + t * p2p3;
    return (1 - t) * p0p1p2 + t * p1p2p3;
}
```

使用 Handles 和 Gizmos 类在 OnDrawGizmos()中绘制三阶贝塞尔曲线，代码如下：

```csharp
//第 6 章/BezierCurvePathExample.cs

private void OnDrawGizmos()
{
    Bezier3Example();
}
private void Bezier3Example()
{
    Gizmos.color = Color.grey;
    Vector3 p0 = Vector3.left * 5f;
    Vector3 p1 = Vector3.left * 2f + Vector3.forward * 2f;
    Vector3 p2 = Vector3.right * 3f + Vector3.back * 4f;
    Vector3 p3 = Vector3.right * 5f;
    Gizmos.DrawLine(p0, p1);
    Gizmos.DrawLine(p1, p2);
    Gizmos.DrawLine(p2, p3);
    Handles.Label(p0, "P0");
    Handles.Label(p1, "P1");
    Handles.Label(p2, "P2");
    Handles.Label(p3, "P3");
    Handles.SphereHandleCap(0, p0, Quaternion.identity,
        0.1f, EventType.Repaint);
    Handles.SphereHandleCap(0, p1, Quaternion.identity,
        0.1f, EventType.Repaint);
    Handles.SphereHandleCap(0, p2, Quaternion.identity,
        0.1f, EventType.Repaint);
    Handles.SphereHandleCap(0, p3, Quaternion.identity,
        0.1f, EventType.Repaint);

    Gizmos.color = Color.green;
    for (int i = 0; i < 100; i++)
    {
        Vector3 curr = BezierCurveUtility.Bezier3(
            p0, p1, p2, p3, i / 100f);
```

```
            Vector3 next = BezierCurveUtility.Bezier3(
                p0, p1, p2, p3, (i + 1) / 100f);
            Gizmos.color = t > (i / 100f)
                ? Color.red : Color.green;
            Gizmos.DrawLine(curr, next);
        }
        Vector3 pt = BezierCurveUtility.Bezier3(p0, p1, p2, p3, t);
        Handles.Label(pt, string.Format("Pt (t = {0})", t));
        Handles.SphereHandleCap(0, pt, Quaternion.identity,
            0.1f, EventType.Repaint);
    }
```

运行程序，结果如图 6-29 所示。

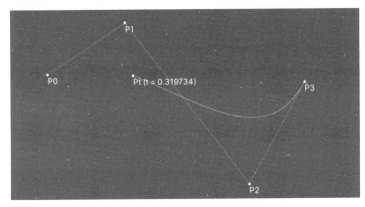

图 6-29 三阶贝塞尔曲线

2. 贝塞尔曲线形成的路径

一条路径由若干路径点形成，前后相邻的两个路径点形成一条贝塞尔曲线，在这个路径点的数据结构中不仅包含坐标点，还包含一个与坐标点形成切线的控制点，代码如下：

```
//第6章/BezierCurvePoint.cs

using System;
using UnityEngine;

[Serializable]
public struct BezierCurvePoint
{
    //<summary>
    //坐标点
    //</summary>
    public Vector3 position;

    //<summary>
    //控制点与坐标点形成切线
```

```
    //</summary>
    public Vector3 tangent;
}
```

将这些路径点存储于一个 List 列表中，并且声明两个变量，一个用于控制路径曲线的平滑程度，另一个用于控制路径是否循环，也就是第 1 个点与最后一个点是否相连而形成闭环，代码如下：

```
//第6章/BezierCurvePath.cs

using UnityEngine;
using System.Collections.Generic;

public class BezierCurvePath : MonoBehaviour
{
    //<summary>
    //段数
    //</summary>
    [Range(1, 100)] public int segments = 10;

    //<summary>
    //是否循环
    //</summary>
    public bool loop;

    //<summary>
    //点集合
    //</summary>
    public List<BezierCurvePoint> points = new List<BezierCurvePoint>(2)
    {
        new BezierCurvePoint()
        {
            position = Vector3.back * 5f,
            tangent = Vector3.back * 5f + Vector3.left * 3f
        },
        new BezierCurvePoint()
        {
            position = Vector3.forward * 5f,
            tangent = Vector3.forward * 5f + Vector3.right * 3f
        }
    };
}
```

当给定一个取值范围为[0,1]的参数时，组件需要返回一个在路径上对应位置的点，此方法的代码如下：

```
//第6章/BezierCurvePath.cs
```

```csharp
//<summary>
//根据归一化位置值获取对应的贝塞尔曲线上的点
//</summary>
//<param name="t">归一化位置值取值范围为[0,1]</param>
//<returns></returns>
public Vector3 EvaluatePosition(float t)
{
    Vector3 retVal = Vector3.zero;
    if (points.Count > 0)
    {
        float max = points.Count - 1 < 1 ? 0
            : (loop ? points.Count : points.Count - 1);
        float standardized = (loop && max > 0) ? ((t %= max)
            + (t < 0 ? max : 0)) : Mathf.Clamp(t, 0, max);
        int rounded = Mathf.RoundToInt(standardized);
        int i1, i2;
        if (Mathf.Abs(standardized - rounded) < Mathf.Epsilon)
            i1 = i2 = (rounded == points.Count) ? 0 : rounded;
        else
        {
            i1 = Mathf.FloorToInt(standardized);
            if (i1 >= points.Count)
            {
                standardized -= max;
                i1 = 0;
            }
            i2 = Mathf.CeilToInt(standardized);
            i2 = i2 >= points.Count ? 0 : i2;
        }
        retVal = i1 == i2 ? points[i1].position
            : BezierCurveUtility.Bezier3(points[i1].position,
            points[i1].position + points[i1].tangent,
            points[i2].position - points[i2].tangent,
            points[i2].position, standardized - i1);
    }
    return retVal;
}
```

3. 路径的绘制与编辑

在组件的 OnDrawGizmos() 方法中将该路径绘制出来，声明一个 Color 类型变量，用于控制路径的绘制颜色，代码如下：

```csharp
//第6章/BezierCurvePath.cs

//<summary>
//路径颜色(Gizmos)
//</summary>
```

```
public Color pathColor = Color.green;

private void OnDrawGizmos()
{
    if (points.Count == 0) return;
    //缓存颜色
    Color cacheColor = Gizmos.color;
    //路径绘制颜色
    Gizmos.color = pathColor;
    //步长
    float step = 1f / segments;
    //缓存的坐标点
    Vector3 lastPos = transform
        .TransformPoint(EvaluatePosition(0f));
    float end = (points.Count - 1 < 1 ? 0
        : (loop ? points.Count : points.Count - 1)) + step * .5f;
    for (float t = step; t <= end; t += step)
    {
        //计算位置
        Vector3 p = transform.TransformPoint(EvaluatePosition(t));
        //绘制曲线
        Gizmos.DrawLine(lastPos, p);
        //记录
        lastPos = p;
    }
    //恢复颜色
    Gizmos.color = cacheColor;
}
```

最后，创建组件的自定义编辑器类，在 OnSceneGUI()方法中使用 Handles 类中的 API 去控制组件中的路径点，并且使用 Undo 中的 RecordObject()方法使编辑器支持撤销、恢复用户的操作，代码如下：

```
//第6章/BezierCurvePathEditor.cs

using UnityEngine;
using UnityEditor;

[CustomEditor(typeof(BezierCurvePath))]
public class BezierCurvePathEditor : Editor
{
    private BezierCurvePath path;
    private const float sphereHandleCapSize = .2f;

    private void OnEnable()
    {
        path = target as BezierCurvePath;
```

```csharp
}

private void OnSceneGUI()
{
    //路径点集合为空
    if (path.points == null || path.points.Count == 0) return;
    //当前选中的工具非移动工具
    if (Tools.current != Tool.Move) return;
    //颜色缓存
    Color cacheColor = Handles.color;
    Handles.color = Color.yellow;
    //遍历路径点集合
    for (int i = 0; i < path.points.Count; i++)
    {
        Vector3 position = DrawPositionHandle(i);
        Vector3 cp = DrawTangentHandle(i);
        //绘制切线
        Handles.DrawDottedLine(position, cp, 1f);
    }
    //恢复颜色
    Handles.color = cacheColor;
}

//路径点操作柄的绘制
private Vector3 DrawPositionHandle(int index)
{
    BezierCurvePoint point = path.points[index];
    //局部转全局坐标
    Vector3 position = path.transform.TransformPoint(point.position);
    //操作柄的旋转类型
    Quaternion rotation = Tools.pivotRotation == PivotRotation.Local
        ? path.transform.rotation : Quaternion.identity;
    //操作柄的大小
    float size = HandleUtility
        .GetHandleSize(position) * sphereHandleCapSize;
    //在该路径点绘制一个球形
    Handles.color = Color.white;
    Handles.SphereHandleCap(0, position,
        rotation, size, EventType.Repaint);
    Handles.Label(position, string.Format("Point{0}", index));
    //检测变更
    EditorGUI.BeginChangeCheck();
    //坐标操作柄
    position = Handles.PositionHandle(position, rotation);
    //变更检测结束,如果发生变更,则更新路径点
    if (EditorGUI.EndChangeCheck())
    {
```

```
        //记录操作
        Undo.RecordObject(path, "Position Changed");
        //全局转局部坐标
        point.position = path.transform.InverseTransformPoint(position);
        //更新路径点
        path.points[index] = point;
    }
    return position;
}

//控制点操作柄的绘制
private Vector3 DrawTangentHandle(int index)
{
    BezierCurvePoint point = path.points[index];
    //局部转全局坐标
    Vector3 cp = path.transform
        .TransformPoint(point.position + point.tangent);
    //操作柄的旋转类型
    Quaternion rotation = Tools.pivotRotation == PivotRotation.Local
        ? path.transform.rotation : Quaternion.identity;
    //操作柄的大小
    float size = HandleUtility
        .GetHandleSize(cp) * sphereHandleCapSize;
    //在该控制点绘制一个球形
    Handles.color = Color.yellow;
    Handles.SphereHandleCap(0, cp, rotation, size, EventType.Repaint);
    //检测变更
    EditorGUI.BeginChangeCheck();
    //坐标操作柄
    cp = Handles.PositionHandle(cp, rotation);
    //变更检测结束，如果发生变更，则更新路径点
    if (EditorGUI.EndChangeCheck())
    {
        //记录操作
        Undo.RecordObject(path, "Control Point Changed");
        //全局转局部坐标
        point.tangent = path.transform
            .InverseTransformPoint(cp) - point.position;
        //更新路径点
        path.points[index] = point;
    }
    return cp;
}
```

最终效果如图 6-30 所示。

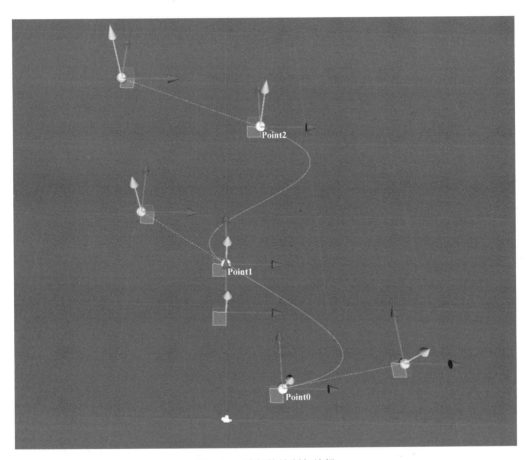

图 6-30　路径的绘制与编辑

4. 如何让物体沿路径移动

在 Update()方法中设置物体的目标坐标和目标朝向可以让物体沿着用户编辑的路径移动，物体的目标坐标位置通过调用路径组件的 EvaluatePosition()方法获取，物体的目标朝向可以通过声明一个 Vector3 类型变量实现，用于每帧缓存物体的坐标点，当前的坐标点减缓存的坐标点就是物体的目标移动方向，代码如下：

```
using UnityEngine;

public class BezierCurvePathAlonger : MonoBehaviour
{
    [SerializeField] private BezierCurvePath path;
    [SerializeField] private float speed = .1f;

    private float normalized = 0f;
    private Vector3 lastPosition;

    private void Update()
```

```
        {
            float t = normalized + speed * Time.deltaTime;
            float max = path.points.Count - 1 < 1 ? 0 : (path.loop
                ? path.points.Count : path.points.Count - 1);
            normalized = (path.loop && max > 0) ? ((t %= max)
                + (t < 0 ? max : 0)) : Mathf.Clamp(t, 0, max);
            transform.position = path.EvaluatePosition(normalized);
            Vector3 forward = transform.position - lastPosition;
            transform.forward = forward != Vector3.zero
                ? forward : transform.forward;
            lastPosition = transform.position;
        }
    }
```

第 7 章 编辑器环境下的数据与资产管理

7.1 EditorPrefs

EditorPrefs 提供了类似于 PlayerPrefs 的数据保存方法，它们的使用方法基本一致，区别在于前者适用于编辑器模式，而后者适用于 Runtime 运行时。EditorPrefs 中包含的方法见表 7-1。

表 7-1 EditorPrefs 中的方法

方法	作用
SetInt()	设置指定键标识的值，值类型为整数型
GetInt()	获取指定键标识对应的整数类型值
SetFloat()	设置指定键标识的值，值类型为浮点数类型
GetFloat()	获取指定键标识对应的浮点数类型值
SetString()	设置指定键标识的值，值类型为字符串类型
GetString()	获取指定键标识对应的字符串类型值
SetBool()	设置指定键标识的值，值类型为布尔类型
GetBool()	获取指定键标识对应的布尔类型值
HasKey()	是否包含指定的键标识
DeleteKey()	删除指定的键标识及其对应值
DeleteAll()	删除所有的键值数据，需谨慎使用

除此之外，可以对支持 JSON 序列化的类型数据进行保存，将其序列化为字符串后通过 SetString() 方法进行保存，获取其值时，通过 GetString() 方法获取字符串数据后将其反序列化为对应的类型，代码如下：

```
//第7章/EditorPrefsUtility.cs

using UnityEngine;
```

```
using UnityEditor;

public class EditorPrefsUtility
{
    public static void SetObject<T>(string key, T t)
    {
        string json = JsonUtility.ToJson(t);
        EditorPrefs.SetString(key, json);
    }
    public static T GetObject<T>(string key)
    {
        if (EditorPrefs.HasKey(key))
        {
            string json = EditorPrefs.GetString(key);
            return JsonUtility.FromJson<T>(json);
        }
        return default;
    }
}
```

与 PlayerPrefs 一样，EditorPrefs 数据存储的位置也在本地计算机的注册表中，前者的存储位置为 HKEY_CURRENT_USER\SOFTWARE\Unity\UnityEditor\CompanyName\ProductName，后者的存储位置在 HKEY_CURRENT_USER\SOFTWARE\Unity Technologies\UnityVersion 中。例如，使用 SetString()方法存储一个字符串类型的键值，然后在注册表中查看存储的数据，如图 7-1 所示，代码如下：

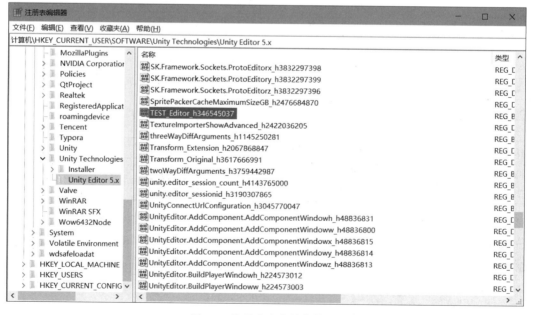

图 7-1　注册表中存储的数据

```
EditorPrefs.SetString("TEST_Editor", "Hello World.");
PlayerPrefs.SetString("TEST_Player", "Hello World.");
```

7.2 AssetDatabase

AssetDatabase 是用于项目工程资产管理的类,仅在编辑器中可用,与其他编辑器类一样,仅适用于放置在 Editor 文件夹中的脚本,使用该类可以对资产进行导入、删除或加载等操作。

需要注意的是,该类中有许多包含表示资产路径的参数的方法,这些路径指的均是相对于项目文件夹的路径,并且都以 Assets/开头。

7.2.1 资产管理

本节介绍 AssetDatabase 类中与资产管理相关的方法及作用,见表 7-2,并通过案例展示相关方法的应用。

表 7-2 AssetDatabase 类中与资产管理相关的方法及作用

方 法	作 用
AddObjectToAsset()	将某个资产对象添加到另一个现有资产中
AssetPathToGUID()	根据资产的路径获取该资产的全局唯一标识符
ClearFolder()	创建一个新的文件夹,方法返回值是新建文件夹的 GUID
ClearLabels()	清除资产的所有标签
Contains()	判断目标对象是否为项目工程中的资产,如果是 Assets 目录中的某个文件,则返回值为 true,如果不是资产,例如场景中的对象或运行时创建的对象,则返回值为 false
CopyAsset()	复制某个资产将其存储在指定的新路径
CreateAsset()	创建一个新资产,并指定资产文件的存储路径
DeleteAsset()	删除指定的资产
ExtractAsset()	在指定的资产中提取子资产,适用于在 fbx 模型资产中提取其中的材质资产,提取成功后返回一个空字符串,否则返回错误消息
FindAssets()	根据筛选器字符串检索资产
ForceReserializeAssets()	强制加载指定资产并将其重新序列化,这样会将所有待定数据更改都刷新到磁盘
GenerateUniqueAssetPath()	为资产创建一个新的唯一路径
GetAssetOrScenePath()	获取资产的存储路径,如果对象是场景中的游戏物体,则该方法返回场景路径
GetAssetPath()	获取资产的路径
GetAssetPathFromTextMetaFilePath()	获取与 meta 文本文件关联的资产文件的路径
GetAvaliableImporterTypes()	根据资产的路径获取所有与资产相关的资产导入器类型

续表

方法	作用
GetCachedIcon()	根据资产路径获取资产的图标
GetImporterOverride()	根据资产的路径获取资产的资产导入器类型
GetLabels()	获取指定资产的所有标签
GetMainAssetTypeAtPath()	根据资产的路径获取主资产对象类型
GetSubFolders()	根据指定文件夹的路径获取所有子目录路径集合
GetTextMetaFilePathFromAssetPath()	根据资产的路径获取与其关联的 meta 文本文件的路径
GetTypeFromPathAndFileID()	根据资产的路径和本地文件标识符获取资产的类型
GUIDFromAssetPath()	根据资产的路径获取资产的全局唯一标识符
GUIDToAssetPath()	根据资产的全局唯一标识符获取资产的路径
ImportAsset()	从指定的路径中导入资产
IsForeignAsset()	确定资产是否为外部资产（导入 Unity 项目中的外部文件）
IsMainAsset()	确定资产是否为 Project 窗口中的主资产
IsMainAssetAtPathLoaded()	确定内存中是否加载了指定资产路径中的主资产
IsMetaFileOpenForEdit()	确定资产关联的 meta 文本文件是否在版本控制系统中打开以供编辑
IsNativeAsset()	确定资产是否为原生资产（Unity 的序列化系统直接生成的文件）
IsOpenForEdit()	确定指定的资产是否在版本控制系统中打开以供编辑
IsSubAsset()	确定资产是否为其他资产的子资产
IsValidFolder()	确定指定的文件夹路径是否存在
LoadAllAssetRepresentationsAtPath()	根据资产的路径加载资产的所有子资产
LoadAllAssetAtPath()	根据资产的路径加载资产的所有资产（一个资产文件可能会包含多个子资产）
LoadAssetAtPath()	根据资产的路径和类型加载资产
LoadMainAssetAtPath()	根据资产的路径加载其主资产
MakeEditable()	在版本控制系统中打开文件以供编辑
MoveAsset()	将一个资产文件或文件夹从一个文件夹移动到另一个文件夹，移动成功后该方法返回空字符串，否则返回错误消息
MoveAssetToTrash()	将资产文件或文件夹移动到垃圾箱中
OpenAsset()	根据资产的类型使用关联的应用程序打开资产
Refresh()	刷新资产，通常在增删资产后调用
RemoveObjectFromAsset()	将资产中的某个资产对象移除
RenameAsset()	为资产文件重命名
SaveAssetIfDirty()	如果资产发生变更，则进行保存
SaveAssets()	对所有未保存的资产变更后进行保存，写入磁盘
SetImporterOverride()	为指定的资产设置特定类型的导入器

续表

方法	作用
SetLabels()	为指定的资产设置标签
SetMainObject()	指定资产文件中的某个对象在下次导入后应成为主对象，资产中的所有其他资产都成为主对象的子项，注意修改的是资产导入器对象，而不是资产本身
TryGetGUIDAndLocalFileIdentifier()	尝试获取资产的全局唯一标识符和本地文件标识符
ValidateMoveAsset()	检查是否可以将资产文件从一个文件夹移动到另一个文件夹，如果可以移动，则返回空字符串，否则返回错误消息

1. 音频数据库

AddObjectToAsset()方法用于将某个资产对象添加到另一个现有资产中，相应地，RemoveObjectFromAsset()方法用于将资产中的某个资产对象移除，代码如下：

```
public static void AddObjectToAsset (Object objectToAdd, string path);
public static void AddObjectToAsset (Object objectToAdd, Object assetObject);
public static void RemoveObjectFromAsset (Object objectToRemove);
```

参数 objectToAdd 表示要添加到现有资产的对象，path 表示现有资产的路径，assetObject 表示现有资产，objectToRemove 表示要移除的资产对象。

本节介绍一个音频数据库的数据容器，它使用一个列表存储所有的音频数据组，每个音频数据组都有一个对应的 ID 和一个存储音频数据的列表，而每个音频数据也都有一个对应的 ID 和 AudioClip 资产，存储音频数据的代码如下：

```
//第7章/AudioData.cs

using System;
using UnityEngine;

//<summary>
//音频数据
//</summary>
[Serializable]
public class AudioData
{
    //<summary>
    //音频数据 ID
    //</summary>
    public int id;
    //<summary>
    //音频剪辑资产
    //</summary>
    public AudioClip clip;
}
```

音频数据组是一个 ScriptableObject 类型资产，代码如下：

```csharp
//第 7 章/AudioGroup.cs

using UnityEngine;
using System.Collections.Generic;

//<summary>
//音频数据组
//</summary>
public class AudioGroup : ScriptableObject
{
    //<summary>
    //音频数据组 ID
    //</summary>
    public int id;
    //<summary>
    //列表存储组中所有的音频数据
    //</summary>
    public List<AudioData> datas = new List<AudioData>(0);

    public AudioClip this[int id]
    {
        get
        {
            int index = datas.FindIndex(m => m.id == id);
            return index != -1 ? datas[index].clip : null;
        }
    }
}
```

音频数据库同样是 ScriptableObject 类型资产，代码如下：

```csharp
//第 7 章/AudioDatabase.cs

using UnityEngine;
using System.Collections.Generic;

//<summary>
//音频数据库
//</summary>
[CreateAssetMenu]
public class AudioDatabase : ScriptableObject
{
    //<summary>
    //列表存储所有的音频数据组
    //</summary>
    public List<AudioGroup> groups = new List<AudioGroup>(0);
```

```csharp
    public AudioGroup this[int id]
    {
        get
        {
            return groups.Find(m => m.id == id);
        }
    }
}
```

音频数据组是从属于音频数据库的资产，因此在创建音频数据组实例时，便可以使用 AssetDatabase 类中的 AddObjectToAsset()方法将其设为音频数据库的子资产。当在音频数据库中删除音频数据组时，通过 RemoveObjectFromAsset()方法将其从中移除。为了实现该功能，为音频数据库创建了自定义编辑器类，代码如下：

```csharp
//第7章/AudioDatabaseEditor.cs

using System;
using System.Linq;
using System.Collections.Generic;

using UnityEngine;
using UnityEditor;

[CustomEditor(typeof(AudioDatabase))]
public class AudioDatabaseEditor : Editor
{
    private AudioDatabase database;
    private Dictionary<AudioData, AudioSource> players;
    private GUIStyle progressStyle;

    private void OnEnable()
    {
        database = target as AudioDatabase;
        players = new Dictionary<AudioData, AudioSource>();
        EditorApplication.update += Update;
    }

    private void OnDestroy()
    {
        EditorApplication.update -= Update;
        foreach (var player in players)
            DestroyImmediate(player.Value.gameObject);
        players.Clear();
    }

    private void Update()
```

```csharp
        {
            Repaint();
            foreach (var player in players)
            {
                if (!player.Value.isPlaying)
                {
                    DestroyImmediate(player.Value.gameObject);
                    players.Remove(player.Key);
                    break;
                }
            }
        }

        public override void OnInspectorGUI()
        {
            if (progressStyle == null)
            {
                progressStyle = new GUIStyle(GUI.skin.label)
                {
                    alignment = TextAnchor.LowerRight,
                    fontSize = 8,
                    fontStyle = FontStyle.Italic
                };
            }
            if (GUILayout.Button("Create New Audio Group"))
            {
                var audioGroup = Activator.CreateInstance<AudioGroup>();
                audioGroup.name = "New Audio Group";
                audioGroup.id = database.groups.Count == 0
                    ? 100 : (database.groups.Count + 100);
                AssetDatabase.AddObjectToAsset(audioGroup, database);
                database.groups.Add(audioGroup);
                AssetDatabase.ImportAsset(
                    AssetDatabase.GetAssetPath(audioGroup));
            }
            EditorGUILayout.Space();
            for (int i = 0; i < database.groups.Count; i++)
            {
                AudioGroup group = database.groups[i];
                OnAudioGroupGUI(group);
                EditorGUILayout.Space();
            }
        }

        private void OnAudioGroupGUI(AudioGroup group)
        {
            GUILayout.BeginHorizontal();
```

```csharp
EditorGUI.BeginChangeCheck();
//Name
string newGroupName = EditorGUILayout.TextField(group.name);
if (newGroupName != group.name)
{
    Undo.RecordObject(database, "Audio Group Name");
    group.name = newGroupName;
}
//ID
var newId = EditorGUILayout.IntField(group.id);
if (newId != group.id)
{
    Undo.RecordObject(database, "Audio Group ID");
    group.id = newId;
}
if (EditorGUI.EndChangeCheck())
{
    serializedObject.ApplyModifiedProperties();
    EditorUtility.SetDirty(database);
    AssetDatabase.SaveAssetIfDirty(database);
}
//删除按钮
if (GUILayout.Button(EditorGUIUtility.IconContent(
    "Toolbar Minus"), GUILayout.Width(20f)))
{
    if (EditorUtility.DisplayDialog(
        "提醒", "是否确认删除该音频数据组", "确认", "取消"))
    {
        Undo.RecordObject(database, "Delete Audio Group");
        database.groups.Remove(group);
        AssetDatabase.RemoveObjectFromAsset(group);
        serializedObject.ApplyModifiedProperties();
        EditorUtility.SetDirty(database);
        AssetDatabase.SaveAssets();
        Repaint();
    }
}
GUILayout.EndHorizontal();
EditorGUILayout.Space();
//Audio Datas
for (int j = 0; j < group.datas.Count; j++)
{
    AudioData data = group.datas[j];
    OnAudioDataGUI(group, data);
}

EditorGUILayout.Space();
```

```csharp
//在以下代码块中绘制了一个矩形区域
//如果将AudioClip资产拖到该区域，则添加一项音频数据
GUILayout.BeginHorizontal();
{
    GUILayout.Label(GUIContent.none, GUILayout.ExpandWidth(true));
    Rect lastRect = GUILayoutUtility.GetLastRect();
    var dropRect = new Rect(lastRect.x + 2f,
        lastRect.y - 2f, 120f, 20f);
    bool containsMouse = dropRect
        .Contains(Event.current.mousePosition);
    if (containsMouse)
    {
        switch (Event.current.type)
        {
            case EventType.DragUpdated:
                bool containsAudioClip = DragAndDrop
                    .objectReferences.OfType<AudioClip>().Any();
                DragAndDrop.visualMode = containsAudioClip
                    ? DragAndDropVisualMode.Copy
                    : DragAndDropVisualMode.Rejected;
                Event.current.Use();
                Repaint();
                break;
            case EventType.DragPerform:
                IEnumerable<AudioClip> audioClips = DragAndDrop
                    .objectReferences.OfType<AudioClip>();
                foreach (var audioClip in audioClips)
                {
                    int index = group.datas.FindIndex(
                        m => m.clip == audioClip);
                    if (index == -1)
                    {
                        Undo.RecordObject(database, "Add Audio Data");
                        int newDataId = group.datas.Count == 0
                            ? 1000 : (group.datas.Count + 1000);
                        group.datas.Add(new AudioData()
                        {
                            id = newDataId,
                            clip = audioClip
                        });
                        serializedObject.ApplyModifiedProperties();
                        EditorUtility.SetDirty(database);
                    }
                }
                Event.current.Use();
                Repaint();
                break;
```

```csharp
            }
        }
        Color color = GUI.color;
        GUI.color = new Color(GUI.color.r, GUI.color.g,
            GUI.color.b, containsMouse ? .9f : .5f);
        GUI.Box(dropRect, "Drop AudioClips Here",
            new GUIStyle(GUI.skin.box) { fontSize = 10 });
        GUI.color = color;
    }
    GUILayout.EndHorizontal();
}

private void OnAudioDataGUI(AudioGroup group, AudioData data)
{
    Color cacheColor = GUI.color;
    GUILayout.BeginHorizontal();
    //绘制音频图标
    GUILayout.Label(EditorGUIUtility
        .IconContent("SceneViewAudio"), GUILayout.Width(20f));
    EditorGUI.BeginChangeCheck();
    //Audio Data ID
    int newDataId = EditorGUILayout
        .IntField(data.id, GUILayout.Width(60f));
    if (newDataId != data.id)
    {
        Undo.RecordObject(database, "Audio Data Id");
        data.id = newDataId;
    }
    //Audio Clip
    AudioClip clip = EditorGUILayout.ObjectField(
        data.clip, typeof(AudioClip), false) as AudioClip;
    if (clip != data.clip)
    {
        Undo.RecordObject(database, "Audio Data Clip");
        data.clip = clip;
    }
    if (EditorGUI.EndChangeCheck())
    {
        serializedObject.ApplyModifiedProperties();
        EditorUtility.SetDirty(database);
    }
    //若该音频正在播放,则计算其播放进度
    string progress = players.ContainsKey(data)
        ? ToTimeFormat(players[data].time) : "00:00:000";
    GUI.color = new Color(GUI.color.r, GUI.color.g,
        GUI.color.b, players.ContainsKey(data) ? .9f : .5f);
    //显示信息:播放进度 / 音频时长 (00:00 / 00:00)
```

```csharp
GUILayout.Label(string.Format("({0} / {1})", progress,
    data.clip != null ? ToTimeFormat(data.clip.length)
    : "00:00:000"), progressStyle, GUILayout.Width(100f));
GUI.color = cacheColor;
//播放按钮
if (GUILayout.Button(EditorGUIUtility
    .IconContent("PlayButton"), GUILayout.Width(20f)))
{
    if (!players.ContainsKey(data))
    {
        //创建一个物体并添加 AudioSource 组件
        var source = EditorUtility.CreateGameObjectWithHideFlags(
            "Audio Player",HideFlags.HideAndDontSave)
            .AddComponent<AudioSource>();
        source.clip = data.clip;
        source.Play();
        players.Add(data, source);
    }
}
//停止播放按钮
if (GUILayout.Button(EditorGUIUtility
    .IconContent("PauseButton"), GUILayout.Width(20f)))
{
    if (players.ContainsKey(data))
    {
        DestroyImmediate(players[data].gameObject);
        players.Remove(data);
    }
}
//删除按钮，单击后删除该项音频数据
if (GUILayout.Button(EditorGUIUtility
    .IconContent("Toolbar Minus"), GUILayout.Width(20f)))
{
    Undo.RecordObject(database, "Delete Audio Data");
    group.datas.Remove(data);
    if (players.ContainsKey(data))
    {
        DestroyImmediate(players[data].gameObject);
        players.Remove(data);
    }
    serializedObject.ApplyModifiedProperties();
    EditorUtility.SetDirty(database);
    Repaint();
}
GUILayout.EndHorizontal();
}
```

```csharp
//将描述转换为mm:ss:fff时间格式字符串
private string ToTimeFormat(float time)
{
    int millSecounds = (int)(time * 1000);
    int minutes = millSecounds / 60000;
    int seconds = millSecounds % 60000 / 1000;
    millSecounds = millSecounds % 60000 % 1000;
    return string.Format("{0:D2}:{1:D2}:{2:D3}",
        minutes, seconds, millSecounds);
}
```

最终效果如图7-2所示。

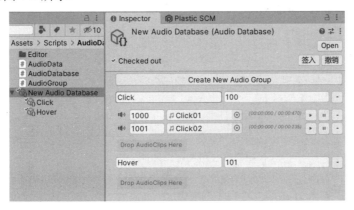

图7-2 音频数据库编辑器

有了音频数据库，可以根据音频数据组和音频数据的ID获取对应的AudioClip资产，例如图7-2中的音频数据库，其有一个ID为100的音频数据组，该组中有一个ID为1000的音频数据，那么便可以通过这两个ID获得对应的音频资产，代码如下：

```csharp
AudioClip clip = database[100]?[1000];
```

2. 资源管理组件

本节介绍一个资源管理组件，该组件用于动态地加载资源和卸载资源，资源加载模式分为编辑器模式、模拟模式、真实模式三种。

真实模式中通过网络请求（UnityWebRequest）从服务器端下载资源，用于真实生产环境。模拟模式是将资源包部署在项目StreamingAssets文件夹中，从该文件夹中下载资源包，模拟资源加载流程的模式。编辑器模式是指在项目开发调试阶段，在编辑器中通过AssetDatabase类中的方法加载资源的模式。

主要用到LoadAssetAtPath()方法，代码如下：

```csharp
//第7章/ResourceComponent.cs

using System;
```

```csharp
using System.IO;
using System.Text;
using System.Collections;
using System.Collections.Generic;
using System.Runtime.Serialization.Formatters.Binary;

using UnityEngine;
using UnityEngine.Networking;
using UnityEngine.SceneManagement;
using Object = UnityEngine.Object;

#if UNITY_EDITOR
using UnityEditor;
using UnityEditor.SceneManagement;
#endif

[DisallowMultipleComponent]
public class ResourceComponent : MonoBehaviour
{
    public enum MODE
    {
        EDITOR,         //编辑器模式
        SIMULATIVE,     //模拟模式(StreamingAssets)
        REALITY,        //真实模式
    }

    [SerializeField]
    MODE mode = MODE.EDITOR;

    [SerializeField]
    string assetBundleUrl = Application.streamingAssetsPath;

    [SerializeField]
    string assetBundleManifestName = "AssetBundles";

    private AssetBundleManifest assetBundleManifest;

    private Dictionary<string, AssetInfo> map
        = new Dictionary<string, AssetInfo>();

    private bool isMapLoading = true;

    private bool isAssetBundleManifestLoading;

    private readonly Dictionary<string, AssetBundle> assetBundlesDic
        = new Dictionary<string, AssetBundle>();
```

```csharp
        private readonly Dictionary<string, Scene> sceneDic
            = new Dictionary<string, Scene>();

        private readonly Dictionary<string, UnityWebRequest> loadingDic
            = new Dictionary<string, UnityWebRequest>();

        private void Start()
        {
#if UNITY_EDITOR
            if (mode != MODE.EDITOR)
                StartCoroutine(LoadAssetsMapAsync());
#else
            StartCoroutine(LoadAssetsMapAsync());
#endif
        }

        private IEnumerator LoadAssetsMapAsync()
        {
            string url = (mode == MODE.REALITY ? assetBundleUrl
                : Application.streamingAssetsPath) + "/map.dat";
            //Debug.Log("资源下载路径: {0}", url);
            using (UnityWebRequest request = UnityWebRequest.Get(url))
            {
#if UNITY_2017_2_OR_NEWER
                yield return request.SendWebRequest();
#else
                yield return request.Send();
#endif
                bool flag = false;
#if UNITY_2020_2_OR_NEWER
                flag = request.result == UnityWebRequest.Result.Success;
#elif UNITY_2017_1_OR_NEWER
                flag = !(request.isNetworkError || request.isHttpError);
#else
                flag = !request.isError;
#endif
                if (flag)
                {
                    string mapPath = Application.persistentDataPath + "/map.dat";
                    File.WriteAllBytes(mapPath, request.downloadHandler.data);
                    //打开文件
                    using (FileStream fs = new FileStream(mapPath, FileMode.Open))
                    {
                        //反序列化
                        BinaryFormatter bf = new BinaryFormatter();
                        var deserialize = bf.Deserialize(fs);
                        if (deserialize != null)
```

```csharp
                    {
                        byte[] buffer = deserialize as byte[];
                        if (buffer != null && buffer.Length > 0)
                        {
                            map = new Dictionary<string, AssetInfo>();
                            string json = Encoding.Default.GetString(buffer);
                            var assetsInfo = JsonUtility
                                .FromJson<AssetsInfo>(json);

                            int counter = 0;
                            for (int i = 0; i < assetsInfo.list.Count; i++)
                            {
                                var info = assetsInfo.list[i];
                                info.name = Path
                                    .GetFileNameWithoutExtension(info.path);
                                map.Add(info.path, info);
                                if (++counter == 100)
                                {
                                    counter = 0;
                                    yield return null;
                                }
                            }
                            isMapLoading = false;
                            Debug.Log("成功加载资源信息");
                        }
                    }
                }
            }
            else
            {
                Debug.LogError(string.Format("请求下载 map.dat 失败：{0} {1}",
                    request.url, request.error));
            }
        }
    }

    private IEnumerator LoadAssetBundleManifestAsync()
    {
        using (UnityWebRequest request = UnityWebRequestAssetBundle
            .GetAssetBundle((mode == MODE.REALITY ? assetBundleUrl
                : Application.streamingAssetsPath) + "/"
                + assetBundleManifestName))
        {
#if UNITY_2017_2_OR_NEWER
            yield return request.SendWebRequest();
#else
```

```csharp
            yield return request.Send();
#endif
            bool flag = false;
#if UNITY_2020_2_OR_NEWER
            flag = request.result == UnityWebRequest.Result.Success;
#elif UNITY_2017_1_OR_NEWER
            flag = !(request.isNetworkError || request.isHttpError);
#else
            flag = !request.isError;
#endif
            if (flag)
            {
                AssetBundle ab = DownloadHandlerAssetBundle
                    .GetContent(request);
                if (ab != null)
                {
                    assetBundleManifest = ab
                        .LoadAsset<AssetBundleManifest>("AssetBundleManifest");
                    isAssetBundleManifestLoading = false;
                }
                else
                {
                    Debug.LogError(string.Format(
                        "下载AssetBundleManifest失败: {0}",
                        request.url));
                }
            }
            else
            {
                Debug.LogError(string.Format(
                    "请求下载AssetBundleManifest失败: {0} {1}",
                    request.url, request.error));
            }
        }
    }

    private IEnumerator LoadAssetBundleAsync(string assetBundleName,
        Action<float> onLoading = null)
    {
        DateTime beginTime = DateTime.Now;

        if (loadingDic.TryGetValue(assetBundleName, out var target))
        {
            yield return null;
            if (target != null)
            {
                while (!target.isDone)
```

```csharp
                    {
                        onLoading?.Invoke(target.downloadProgress);
                        yield return null;
                    }
                }
                yield return new WaitUntil(() =>
                    !loadingDic.ContainsKey(assetBundleName));
            }
            else
            {
                using (UnityWebRequest request = UnityWebRequestAssetBundle
                    .GetAssetBundle((mode == MODE.REALITY ? assetBundleUrl
                    : Application.streamingAssetsPath) + "/" + assetBundleName))
                {
                    loadingDic.Add(assetBundleName, request);
#if UNITY_2017_2_OR_NEWER
                    yield return request.SendWebRequest();
#else
                    yield return request.Send();
#endif
                    while (!request.isDone)
                    {
                        onLoading?.Invoke(request.downloadProgress);
                        yield return null;
                    }

                    bool flag = false;
#if UNITY_2020_2_OR_NEWER
                    flag = request.result == UnityWebRequest.Result.Success;
#elif UNITY_2017_1_OR_NEWER
                    flag = !(request.isNetworkError || request.isHttpError);
#else
                    flag = !request.isError;
#endif
                    if (flag)
                    {
                        AssetBundle ab = DownloadHandlerAssetBundle
                            .GetContent(request);
                        if (ab != null)
                        {
                            assetBundlesDic.Add(assetBundleName, ab);
                            Debug.Log(string.Format("于{0}发起下载AssetBundle 请求{1}" +
                                "于{2}下载完成,耗时{3}毫秒({4}秒)",
                                beginTime.ToString("T"), request.url,
                                DateTime.Now.ToString("T"),
                                (DateTime.Now - beginTime).TotalMilliseconds,
                                (DateTime.Now - beginTime).TotalSeconds));
```

```csharp
            }
            else
            {
                Debug.LogError(string.Format(
                    "下载AssetBundle失败：{0}", request.url));
            }
        }
        else
        {
            Debug.LogError(string.Format(
                "请求下载AssetBundle失败：{0} {1}",
                request.url, request.error));
        }
        yield return null;
        loadingDic.Remove(assetBundleName);
    }
}


IEnumerator LoadAssetBundleDependeciesAsync(string assetBundleName)
{
    if (assetBundleManifest == null)
    {
        if (isAssetBundleManifestLoading)
        {
            yield return new WaitUntil(()
                => assetBundleManifest != null);
        }
        else
        {
            isAssetBundleManifestLoading = true;
            yield return LoadAssetBundleManifestAsync();
        }
    }

    string[] dependencies = assetBundleManifest
        .GetAllDependencies(assetBundleName);
    for (int i = 0; i < dependencies.Length; i++)
    {
        string dep = dependencies[i];
        if (!assetBundlesDic.ContainsKey(dep))
        {
            yield return LoadAssetBundleAsync(dep);
        }
    }
}
```

```csharp
    IEnumerator LoadAssetAsyncCoroutine<T>(string assetPath,
        Action<bool, T> onCompleted, Action<float> onLoading) where T : Object
    {
        Object asset = null;

#if UNITY_EDITOR
        if (mode == MODE.EDITOR)
        {
            onLoading?.Invoke(1);
            yield return null;

            asset = AssetDatabase.LoadAssetAtPath<T>(assetPath);
            if (asset == null)
            {
                Debug.LogError(string.Format("加载资源失败：{0}", assetPath));
            }
        }
        else
        {
            if (isMapLoading)
            {
                yield return new WaitUntil(() => isMapLoading == false);
            }

            if (!map.TryGetValue(assetPath, out var assetInfo))
            {
                Debug.LogError(string.Format("加载资源失败：{0}", assetPath));
                yield break;
            }

            yield return LoadAssetBundleDependeciesAsync(assetInfo.abName);

            if (!assetBundlesDic.ContainsKey(assetInfo.abName))
            {
                yield return LoadAssetBundleAsync(
                    assetInfo.abName, onLoading);
            }
            else
            {
                onLoading?.Invoke(1);
                yield return null;
            }
            asset = assetBundlesDic[assetInfo.abName]
                .LoadAsset<T>(assetInfo.name);
            if (asset == null)
            {
                Debug.LogError(string.Format("加载资源失败：{0} {1}",
```

```csharp
                    assetInfo.abName, asset.name));
            }
        }
#else
        if (isMapLoading)
        {
            yield return new WaitUntil(() => isMapLoading == false);
        }

        if (!map.TryGetValue(assetPath, out var assetInfo))
        {
            Debug.LogError(string.Format("加载资源失败：{0}", assetPath));
            yield break;
        }

        yield return LoadAssetBundleDependeciesAsync(assetInfo.abName);

        if (!assetBundlesDic.ContainsKey(assetInfo.abName))
        {
            yield return LoadAssetBundleAsync(
                assetInfo.abName, onLoading);
        }
        else
        {
            onLoading?.Invoke(1);
            yield return null;
        }
        asset = assetBundlesDic[assetInfo.abName]
            .LoadAsset<T>(assetInfo.name);
        if (asset == null)
        {
            Debug.LogError(string.Format("加载资源失败：{0} {1}",
                assetInfo.abName, asset.name));
        }
#endif
        if (asset != null)
        {
            onCompleted?.Invoke(true, asset as T);
        }
        else
        {
            onCompleted?.Invoke(false, null);
        }
    }

    private IEnumerator LoadSceneAsyncCoroutine(string sceneAssetPath,
```

```csharp
            Action<bool> onCompleted, Action<float> onLoading)
    {
#if UNITY_EDITOR
        if (mode == MODE.EDITOR)
        {
            if (sceneDic.ContainsKey(sceneAssetPath))
            {
                Debug.LogWarning(string.Format("场景{0}已加载",
                    sceneAssetPath));
                onCompleted?.Invoke(false);
                yield break;
            }

            sceneDic.Add(sceneAssetPath, new Scene());
            AsyncOperation asyncOperation = EditorSceneManager
                .LoadSceneAsyncInPlayMode(sceneAssetPath,
                new LoadSceneParameters()
                {
                    loadSceneMode = LoadSceneMode.Additive,
                    localPhysicsMode = LocalPhysicsMode.None
                });
            while (!asyncOperation.isDone)
            {
                onLoading?.Invoke(asyncOperation.progress);
                yield return null;
            }
            onLoading?.Invoke(1f);
            Scene scene = EditorSceneManager
                .GetSceneByPath(sceneAssetPath);
            sceneDic[sceneAssetPath] = scene;
        }
        else
        {
            if (isMapLoading)
            {
                yield return new WaitUntil(() => isMapLoading == false);
            }
            if (!map.TryGetValue(sceneAssetPath, out var assetInfo))
            {
                Debug.LogError(string.Format("加载场景失败：{0}",
                    sceneAssetPath));
                yield break;
            }
            if (sceneDic.ContainsKey(assetInfo.name))
            {
                Debug.LogWarning(string.Format("场景{0}已加载",
                    sceneAssetPath));
```

```csharp
                onCompleted?.Invoke(false);
                yield break;
            }

            yield return LoadAssetBundleDependeciesAsync(assetInfo.abName);

            Scene scene = SceneManager.GetSceneByPath(sceneAssetPath);
            sceneDic.Add(assetInfo.name, scene);
            if (!assetBundlesDic.ContainsKey(assetInfo.abName))
            {
                yield return LoadAssetBundleAsync(
                    assetInfo.abName, onLoading);
            }
            AsyncOperation asyncOperation = SceneManager
                .LoadSceneAsync(assetInfo.name, LoadSceneMode.Additive);
            while (!asyncOperation.isDone)
            {
                onLoading?.Invoke(asyncOperation.progress);
                yield return null;
            }
            onLoading?.Invoke(1f);
        }
#else
            if (isMapLoading)
            {
                yield return new WaitUntil(() => isMapLoading == false);
            }
            if (!map.TryGetValue(sceneAssetPath, out var assetInfo))
            {
                Debug.LogError(string.Format("加载场景失败：{0}",
                    sceneAssetPath));
                yield break;
            }
            if (sceneDic.ContainsKey(assetInfo.name))
            {
                Debug.LogWarning(string.Format("场景{0}已加载",
                    sceneAssetPath));
                onCompleted?.Invoke(false);
                yield break;
            }

            yield return LoadAssetBundleDependeciesAsync(assetInfo.abName);

            Scene scene = SceneManager.GetSceneByPath(sceneAssetPath);
            sceneDic.Add(assetInfo.name, scene);
            if (!assetBundlesDic.ContainsKey(assetInfo.abName))
            {
```

```
                yield return LoadAssetBundleAsync(
                    assetInfo.abName, onLoading);
            }
            AsyncOperation asyncOperation = SceneManager
                .LoadSceneAsync(assetInfo.name, LoadSceneMode.Additive);
            while (!asyncOperation.isDone)
            {
                onLoading?.Invoke(asyncOperation.progress);
                yield return null;
            }
            onLoading?.Invoke(1f);
#endif
            onCompleted?.Invoke(true);
        }

        //<summary>
        //异步加载资产
        //</summary>
        //<typeparam name="T">资产类型</typeparam>
        //<param name="assetPath">资产路径</param>
        //<param name="onCompleted">加载完成回调事件</param>
        //<param name="onLoading">加载进度回调事件</param>
        public void LoadAssetAsync<T>(string assetPath,
            Action<bool, T> onCompleted = null,
            Action<float> onLoading = null) where T : Object
        {
            StartCoroutine(LoadAssetAsyncCoroutine(
                assetPath, onCompleted, onLoading));
        }

        //<summary>
        //异步加载场景
        //</summary>
        //<param name="sceneAssetPath">场景资产路径</param>
        //<param name="onCompleted">加载完成回调事件</param>
        //<param name="onLoading">加载进度回调事件</param>
        public void LoadSceneAsync(string sceneAssetPath,
            Action<bool> onCompleted = null,
            Action<float> onLoading = null)
        {
            StartCoroutine(LoadSceneAsyncCoroutine(
                sceneAssetPath, onCompleted, onLoading));
        }

        //<summary>
        //卸载资产
        //</summary>
```

```csharp
//<param name="assetPath">资产路径</param>
//<param name="unloadAllLoadedObjects">是否卸载相关实例化对象</param>
public void UnloadAsset(string assetPath,
    bool unloadAllLoadedObjects = false)
{
    if (map.TryGetValue(assetPath, out var assetInfo))
    {
        if (assetBundlesDic.ContainsKey(assetInfo.abName))
        {
            assetBundlesDic[assetInfo.abName]
                .Unload(unloadAllLoadedObjects);
            assetBundlesDic.Remove(assetInfo.abName);
        }
    }
}

//<summary>
//卸载所有资产
//</summary>
//<param name="unloadAllLoadedObjects">是否卸载相关实例化对象</param>
public void UnloadAllAsset(bool unloadAllLoadedObjects = false)
{
    foreach (var kv in assetBundlesDic)
    {
        kv.Value.Unload(unloadAllLoadedObjects);
    }
    assetBundlesDic.Clear();
    AssetBundle.UnloadAllAssetBundles(unloadAllLoadedObjects);
}

//<summary>
//卸载场景
//</summary>
//<param name="sceneAssetPath">场景资产路径</param>
//<returns>true: 卸载成功; false: 卸载失败</returns>
public bool UnloadScene(string sceneAssetPath)
{
#if UNITY_EDITOR
    if (mode == MODE.EDITOR)
    {
        if (sceneDic.TryGetValue(sceneAssetPath, out Scene scene))
        {
            sceneDic.Remove(sceneAssetPath);
            EditorSceneManager.UnloadSceneAsync(scene);
            return true;
        }
        return false;
    }
```

```csharp
            }
            else
            {
                if (map.TryGetValue(sceneAssetPath, out var assetInfo))
                {
                    if (sceneDic.ContainsKey(assetInfo.name))
                    {
                        sceneDic.Remove(assetInfo.name);
                        SceneManager.UnloadSceneAsync(assetInfo.name);
                        return true;
                    }
                }
                return false;
            }
#else
            if (map.TryGetValue(sceneAssetPath, out var assetInfo))
            {
                if (sceneDic.ContainsKey(assetInfo.name))
                {
                    sceneDic.Remove(assetInfo.name);
                    SceneManager.UnloadSceneAsync(assetInfo.name);
                    return true;
                }
            }
            return false;
#endif
        }
}
```

7.2.2 AssetBundle 管理

AssetDatabase 类中与 AssetBundle 相关的方法见表 7-3。

表 7-3 与 AssetBundle 相关的方法

方　　法	作　　用
GetAllAssetBundleNames()	获取所有 AssetBundle 包名称
GetAssetBundleDependencies()	获取指定 AssetBundle 包的依赖项
GetAssetDependencyHash()	获取资产所有依赖项的哈希值
GetAssetPathsFromAssetBundle()	获取指定的 AssetBundle 包中所有资产的路径
GetAssetPathsFromAssetBundleAndAssetName()	获取指定的 AssetBundle 包中包含指定资产名称的所有资产路径
GetDependencies()	获取指定资产的依赖项，此方法返回的是依赖项的路径集合

续表

方　法	作　用
GetImplicitAssetBundleName()	获取指定资产所属的 AssetBundle 包名称
GetImplicitAssetBundleVariantName()	获取指定资产所属的 AssetBundle 包变体的名称
GetUnusedAssetBundleNames()	获取所有未使用的 AssetBundle 名称
RemoveAssetBundleName()	删除 AssetBundle 名称
RemoveUnusedAssetBundleNames()	删除所有未使用的 AssetBundle 名称

本节通过一个用于配置 AssetBundle 资源包的工具展示相关方法的应用，如图 7-3 所示。工具窗口被分为左右两部分。左侧部分通过滚动视图列举了所有的 AssetBundle 名称，开发者可以将资产拖曳到该区域中，拖曳完成后工具会自动根据资产的名称创建 AssetBundle，列表下方绘制了当前选中的资源包的详细信息。右侧部分列举的是当前选中的 AssetBundle 中的所有资产，列表下方绘制了当前选中资产的详细信息。

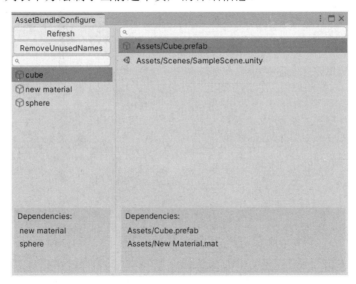

图 7-3　AssetBundle 配置工具

当拖曳资产产生成新的资源包时，为了保证资源包名称的唯一性，需要先检查是否已经存在指定的名称。如果已存在指定的名称，则在名称后附加数字 1 再次检查，持续递增数字并再次检查，直到资源包名称是唯一的，此方法的实现代码如下：

```
//第7章/AssetBundleUtility.cs

using System;
using System.Text.RegularExpressions;

using UnityEngine;
using UnityEditor;
```

```csharp
using Object = UnityEngine.Object;

public class AssetBundleUtility
{
    //<summary>
    //以递归方式获取一个唯一的AssetBundle名称
    //检查是否存在具有相同名称的AssetBundle
    //如果有，则方法会将数字1附加到名称后再次检查
    //持续递增数字并再次检查，直到名称是唯一的
    //</summary>
    //<param name="assetBundleName">AssetBundle名称</param>
    //<returns></returns>
    public static string GetUniqueAssetBundleNameRecursive(
        string assetBundleName)
    {
        //获取当前所有的AssetBundle名称
        string[] assetBundleNames = AssetDatabase.GetAllAssetBundleNames();
        //查找当前是否已经包含该名称
        int index = Array.FindIndex(assetBundleNames,
            m => m == assetBundleName.ToLower());
        //已经包含
        if (index != -1)
        {
            string target = assetBundleNames[index];
            //匹配字符串末尾的数字内容
            string numberStr = Regex.Match(target, @"\d+$").Value;
            //数字内容不为空
            if (!string.IsNullOrEmpty(numberStr))
            {
                //尝试类型转换
                int.TryParse(numberStr, out int number);
                //截掉末尾数字内容，字符截取长度
                int subLength = target.Length - numberStr.Length;
                //拼接自增后的数字以形成新的名称
                string newName = target.Substring(target.Length -
                    subLength - 1, subLength) + (++number);
                //递归获取，直到名称是当前唯一的
                return GetUniqueAssetBundleNameRecursive(newName);
            }
            //数字内容为空
            else
                return GetUniqueAssetBundleNameRecursive(
                    assetBundleName + 1);
        }
        //当前不包含该名称，已经是唯一的名称
        else return assetBundleName;
    }
}
```

为资产设置 AssetBundle 需要先根据资产的路径获取对应的资产导入器，然后修改资产导入器中的 assetBundleName 变量即可，此方法的实现代码如下：

```csharp
//第 7 章/AssetBundleUtility.cs

//<summary>
//为资产创建 AssetBundle
//</summary>
//<param name="obj">资产</param>
//<returns>如果创建成功，则返回值为 true，否则返回值为 false</returns>
public static bool CreateAssetBundle4Object(Object obj)
{
    //根据资产获取资产的路径
    string assetPath = AssetDatabase.GetAssetPath(obj);
    //根据资产路径获取资产导入器
    AssetImporter importer = AssetImporter.GetAtPath(assetPath);
    if (importer == null) return false;
    //为资产设置 AssetBundle 名称
    importer.assetBundleName =
        GetUniqueAssetBundleNameRecursive(obj.name);
    return true;
}
```

当需要删除指定的 AssetBundle 时，首先遍历该 AssetBundle 中所有的资产，将这些资产的 AssetBundle 名称清空，然后移除该 AssetBundle 名称，代码如下：

```csharp
//第 7 章/AssetBundleUtility.cs

//<summary>
//删除 AssetBundle 名称
//</summary>
//<param name="assetBundleName">AssetBundle 名称</param>
public static void DeleteAssetBundleName(string assetBundleName)
{
    //根据 AssetBundle 名称获取其中的资产路径集合
    string[] assetPaths = AssetDatabase
        .GetAssetPathsFromAssetBundle(assetBundleName);
    for (int i = 0; i < assetPaths.Length; i++)
    {
        //根据资产路径获取资产导入器
        AssetImporter importer = AssetImporter.GetAtPath(assetPaths[i]);
        if (importer == null) continue;
        //清空资产的 AssetBundle 名称
        importer.assetBundleName = null;
    }
    //最终移除 AssetBundle 名称
```

```
        AssetDatabase.RemoveAssetBundleName(assetBundleName, true);
}
```

拖曳的资产可以通过 DragAndDrop 类中的静态变量 objectReferences 获取，窗口工具类的代码如下：

```
//第7章/AssetBundleConfigure.cs

using System;
using System.Linq;
using System.Collections.Generic;

using UnityEngine;
using UnityEditor;
using Object = UnityEngine.Object;

public class AssetBundleConfigure : EditorWindow
{
    [MenuItem("Example/Resource/AssetBundle Configure")]
    public static void Open()
    {
        GetWindow<AssetBundleConfigure>().Show();
    }

    private Vector2 lScrollPosition, rScrollPosition;
    private Vector2 abDetailScrollPosition;
    private Vector2 assetDetailScrollPosition;
    //分割线宽度
    private const float splitterWidth = 2f;
    //分割线位置
    private float splitterPos;
    private Rect splitterRect;
    //是否正在拖曳分割线
    private bool isDragging;

    //AssetBundle 名称集合
    private string[] assetBundleNames;
    //<AssetBundle 名称, Assets 路径集合>
    private Dictionary<string, string[]> map;
    //当前选中的 AssetBundle 名称
    private string selectedAssetBundleName;
    //当前选中的 Asset 路径
    private string selectedAssetPath;

    //检索 AssetBundle
    private string searchAssetBundle;
    //检索 Asset 路径
```

```csharp
private string searchAssetPath;

private void OnEnable()
{
    splitterPos = position.width * .5f;
    Init();
}

private void OnDisable()
{
    map = null;
    searchAssetBundle = null;
    selectedAssetBundleName = null;
}

private void OnGUI()
{
    GUILayout.BeginHorizontal();
    lScrollPosition = GUILayout.BeginScrollView(
        lScrollPosition,
        GUILayout.Width(splitterPos),
        GUILayout.MaxWidth(splitterPos),
        GUILayout.MinWidth(splitterPos));
    OnLeftGUI();
    GUILayout.EndScrollView();

    //分割线
    GUILayout.Box(string.Empty,
        GUILayout.Width(splitterWidth),
        GUILayout.MaxWidth(splitterWidth),
        GUILayout.MinWidth(splitterWidth),
        GUILayout.ExpandHeight(true));
    splitterRect = GUILayoutUtility.GetLastRect();

    rScrollPosition = GUILayout.BeginScrollView(
        rScrollPosition, GUILayout.ExpandWidth(true));
    OnRightGUI();
    GUILayout.EndScrollView();
    GUILayout.EndHorizontal();

    if (Event.current != null)
    {
        //光标
        EditorGUIUtility.AddCursorRect(splitterRect,
            MouseCursor.ResizeHorizontal);
        switch (Event.current.rawType)
        {
```

```csharp
            //开始拖曳分割线
            case EventType.MouseDown:
                isDragging = splitterRect.Contains(
                    Event.current.mousePosition);
                break;
            case EventType.MouseDrag:
                if (isDragging)
                {
                    splitterPos += Event.current.delta.x;
                    //限制其最大值和最小值
                    splitterPos = Mathf.Clamp(splitterPos,
                        position.width * .2f, position.width * .8f);
                    Repaint();
                }
                break;
            //结束拖曳分割线
            case EventType.MouseUp:
                if (isDragging)
                    isDragging = false;
                break;
        }
    }
}

private void Init(bool reselect = true)
{
    if (reselect)
    {
        selectedAssetBundleName = null;
        selectedAssetPath = null;
    }
    //获取所有 AssetBundle 名称
    assetBundleNames = AssetDatabase.GetAllAssetBundleNames();
    //初始化 map 字典
    map = new Dictionary<string, string[]>();
    for (int i = 0; i < assetBundleNames.Length; i++)
    {
        map.Add(assetBundleNames[i], AssetDatabase
            .GetAssetPathsFromAssetBundle(assetBundleNames[i]));
    }
}

private void OnLeftGUI()
{
    //刷新,重新加载 AssetBundle 信息
    if (GUILayout.Button("Refresh"))
    {
```

```csharp
    Init();
    Repaint();
}
//移除未使用的AssetBundle名称
if (GUILayout.Button("RemoveUnusedNames"))
{
    AssetDatabase.RemoveUnusedAssetBundleNames();
    Init();
    Repaint();
}
//检索输入框
searchAssetBundle = GUILayout.TextField(
    searchAssetBundle, EditorStyles.toolbarSearchField);
Rect lastRect = GUILayoutUtility.GetLastRect();
//当单击鼠标且鼠标位置不在输入框中时取消控件的聚焦
if (Event.current.type == EventType.MouseDown
    && !lastRect.Contains(Event.current.mousePosition))
{
    GUI.FocusControl(null);
    Repaint();
}

//列表区域
Rect listRect = new Rect(0f, lastRect.y + 20f,
    lastRect.width, position.height - lastRect.y - 25f);
//如果将资产拖曳到列表区域，则为资产创建AssetBundle
if (DragObjects2RectCheck(listRect, out Object[] objects))
{
    bool flag = false;
    for (int i = 0; i < objects.Length; i++)
    {
        if (AssetBundleUtility
            .CreateAssetBundle4Object(objects[i]))
            flag = true;
    }
    //创建了新的AssetBundle，刷新
    if (flag)
    {
        Init();
        Repaint();
    }
}

if (assetBundleNames.Length == 0) return;
for (int i = 0; i < assetBundleNames.Length; i++)
{
    string assetBundleName = assetBundleNames[i];
```

```csharp
            if (!string.IsNullOrEmpty(searchAssetBundle)
                && !assetBundleName.ToLower()
                    .Contains(searchAssetBundle.ToLower()))
                continue;
            GUILayout.BeginHorizontal(selectedAssetBundleName
                == assetBundleName ? "MeTransitionSelectHead"
                : "ProjectBrowserHeaderBgTop",
                GUILayout.Height(20f));
            GUILayout.Label(EditorGUIUtility.TrTextContentWithIcon(
                assetBundleName, "GameObject Icon"), GUILayout.Height(18f));
            GUILayout.EndHorizontal();
            //鼠标单击事件
            if (Event.current.type == EventType.MouseDown
                && GUILayoutUtility.GetLastRect()
                    .Contains(Event.current.mousePosition))
            {
                //如果左键被单击，则选中该项
                if (Event.current.button == 0)
                {
                    selectedAssetBundleName = assetBundleName;
                    Repaint();
                }
                //如果右键被单击，则弹出菜单
                if (Event.current.button == 1)
                {
                    GenericMenu gm = new GenericMenu();
                    //删除 AssetBundle
                    gm.AddItem(new GUIContent("Delete AssetBundle"), false, () =>
                    {
                        //二次确认弹窗
                        if (EditorUtility.DisplayDialog("提醒",
                            string.Format("是否确认删除{0}？",
                                assetBundleName), "确认", "取消"))
                        {
                            AssetBundleUtility.DeleteAssetBundleName(
                                assetBundleName);
                            Init();
                            Repaint();
                        }
                    });
                    gm.ShowAsContext();
                }
            }
        }
    }

    GUILayout.FlexibleSpace();
    GUILayout.BeginVertical(GUI.skin.box, GUILayout.Height(100f),
```

```csharp
            GUILayout.ExpandWidth(true));
        string[] dependencies = AssetDatabase.GetAssetBundleDependencies(
            selectedAssetBundleName, true);
        GUILayout.Label("Dependencies:");
        abDetailScrollPosition = GUILayout.BeginScrollView(
            abDetailScrollPosition);
        for (int i = 0; i < dependencies.Length; i++)
        {
            GUILayout.Label(dependencies[i]);
        }
        GUILayout.EndScrollView();
        GUILayout.EndVertical();
    }

    private void OnRightGUI()
    {
        //检索输入框
        searchAssetPath = GUILayout.TextField(searchAssetPath,
            EditorStyles.toolbarSearchField);
        Rect lastRect = GUILayoutUtility.GetLastRect();
        //当单击鼠标且鼠标位置不在输入框中时取消控件的聚焦
        if (Event.current.type == EventType.MouseDown
            && !lastRect.Contains(Event.current.mousePosition))
        {
            GUI.FocusControl(null);
            Repaint();
        }
        if (selectedAssetBundleName == null) return;

        //资产拖曳区域
        Rect dragRect = new Rect(lastRect.x, lastRect.y + 20f,
            lastRect.width - 5f, position.height - lastRect.y - 25f);
        //如果将资产拖曳到该区域，则为这些资产设置AssetBundle名称
        if (DragObjects2RectCheck(dragRect, out Object[] objects))
        {
            for (int i = 0; i < objects.Length; i++)
            {
                string assetPath = AssetDatabase.GetAssetPath(objects[i]);
                AssetImporter importer = AssetImporter.GetAtPath(assetPath);
                if (importer != null)
                    importer.assetBundleName = selectedAssetBundleName;
            }
            Init(false);
            Repaint();
            return;
        }
```

```csharp
//该AssetBundle中的资产路径集合
string[] assetPaths = map[selectedAssetBundleName];
if (assetPaths.Length == 0) return;
for (int i = 0; i < assetPaths.Length; i++)
{
    string assetPath = assetPaths[i];
    //当前项是否符合检索内容
    if (!string.IsNullOrEmpty(searchAssetPath)
        && !assetPath.ToLower().Contains(
            searchAssetPath.ToLower()))
        continue;
    GUILayout.BeginHorizontal(selectedAssetPath == assetPath
        ? "MeTransitionSelectHead"
        : "ProjectBrowserHeaderBgTop",
        GUILayout.Height(20f));
    Type type = AssetDatabase.GetMainAssetTypeAtPath(assetPath);
    Texture texture = AssetPreview.GetMiniTypeThumbnail(type);
    GUILayout.Label(texture, GUILayout.Width(18f),
        GUILayout.Height(18f));
    GUILayout.Label(assetPaths[i]);
    GUILayout.EndHorizontal();
    //鼠标单击事件
    if (Event.current.type == EventType.MouseDown
        && GUILayoutUtility.GetLastRect()
            .Contains(Event.current.mousePosition))
    {
        //如果左键被单击,则选中该项
        if (Event.current.button == 0)
        {
            selectedAssetPath = assetPath;
            Repaint();
            EditorGUIUtility.PingObject(
                AssetDatabase.LoadMainAssetAtPath(
                    selectedAssetPath));
        }
        //如果右键被单击,则弹出菜单
        else if (Event.current.button == 1)
        {
            GenericMenu gm = new GenericMenu();
            //删除
            gm.AddItem(new GUIContent("Delete"), false, () =>
            {
                //根据资产路径获取其资产导入器
                AssetImporter importer = AssetImporter
                    .GetAtPath(assetPath);
                //清空AssetBundle名称
                if (importer != null)
```

```csharp
                    importer.assetBundleName = null;
                    Init(false);
                    Repaint();
                });
                gm.ShowAsContext();
            }
        }
    }

    GUILayout.FlexibleSpace();
    GUILayout.BeginVertical(GUI.skin.box, GUILayout.Height(100f)
        , GUILayout.ExpandWidth(true));
    string[] dependencies = AssetDatabase.GetDependencies(
        selectedAssetPath, true);
    GUILayout.Label("Dependencies:");
    assetDetailScrollPosition = GUILayout.BeginScrollView(
        assetDetailScrollPosition);
    for (int i = 0; i < dependencies.Length; i++)
    {
        GUILayout.Label(dependencies[i]);
    }
    GUILayout.EndScrollView();
    GUILayout.EndVertical();
}

//是否将资产拖曳到矩形区域中
private bool DragObjects2RectCheck(Rect rect, out Object[] objects)
{
    objects = null;
    //鼠标是否在矩形区域中
    if (rect.Contains(Event.current.mousePosition))
    {
        switch (Event.current.type)
        {
            case EventType.DragUpdated:
                //是否拖曳了资产
                bool containsObjects = DragAndDrop
                    .objectReferences.Any();
                DragAndDrop.visualMode = containsObjects
                    ? DragAndDropVisualMode.Copy
                    : DragAndDropVisualMode.Rejected;
                Event.current.Use();
                Repaint();
                return false;
            case EventType.DragPerform:
                //拖曳的资产
                objects = DragAndDrop.objectReferences;
```

```
                    Event.current.Use();
                    Repaint();
                    return true;
            }
        }
        return false;
    }
}
```

7.2.3 Custom Package 管理

后缀为.unitypackage 的文件是 Unity 中的一种文件格式，用于打包、共享项目中的资产和脚本文件。它可以包含各种类型的资产，如场景、模型、纹理、音频、脚本等。通过打包这些资产，可以方便地在不同的项目之间进行共享和传递。

在编辑器中导入和导出该格式的文件分别通过选择 Assets→Import Package→Custom Package 和 Assets→Export Package 完成，而在脚本中通过代码导入和导出该格式的文件需要调用 Asset Database 类中的 ImportPackage()和 ExportPackage()方法。

ImportPackage()方法根据.unitypackage 文件的路径导入其中的内容，代码如下，参数 packagePath 表示文件的路径，interactive 表示在导入时是否打开导入对话框。

```
public static void ImportPackage (string packagePath, bool interactive);
```

在导入对话框中可以选择要导入的内容，如图 7-4 所示，若 interactive 参数为 false，则不会打开导入对话框，包中的所有内容都会被直接导入工程中。

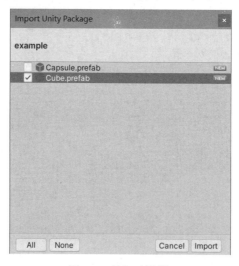

图 7-4 导入对话框

ExportPackage()方法用于将指定的资产导出到一个.unitypackage 文件中，代码如下，参数 assetPathName 表示资产的路径，assetPathNames 表示资产路径集合，当导出多个资产时使用此参数，fileName 表示导出的.unitypackage 文件的路径，flags 表示导出选项。

```
public static void ExportPackage (string assetPathName, string fileName);
public static void ExportPackage (string assetPathName, string fileName,
ExportPackageOptions flags);
public static void ExportPackage (string[] assetPathNames, string fileName,
ExportPackageOptions flags= ExportPackageOptions.Default);
```

7.3 ScriptableObject

ScriptableObject 是用于存储数据的数据容器，与 MonoBehaviour 一样，它继承自 Object 类，但是它不能像 MonoBehaviour 一样挂载于游戏物体上。它作为一种资产被保存为.asset 文件，在编辑器中通常被作为配置文件使用。

实现 ScriptableObject 的派生类，并为其使用 CreateAssetMenu 特性进行标记，该特性用于为创建该类型的实例提供菜单。通过变量 fileName 可以指定新建该类型的实例所使用的文件名，默认以"New +类型名称"命名。通过变量 menuName 可以指定在 Assets/Create 菜单中创建该类型实例的菜单路径，默认为类型名称。通过变量 order 可以指定在 Assets/Create 菜单中创建该类型实例的菜单路径的显示顺序。

示例代码如下：

```
//第7章/ExampleScriptableObject.cs

using UnityEngine;

[CreateAssetMenu(fileName = "New Example SO",
    menuName = "ExampleSO", order = 0)]
public class ExampleScriptableObject : ScriptableObject
{
    public string title;
    public string description;
}
```

选择 Assets→Create→Example SO 命令创建一个该类型的实例，结果如图 7-5 所示。

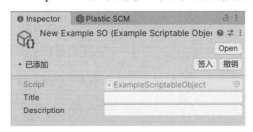

图 7-5　ScriptableObject 资产

开发者可以在 Project Settings 或 Preferences 中添加设置界面，在其中使用 ScriptableObject

的派生类型资产作为某种全局配置。以 OpenXR 为例,它在 Project Settings 中添加了一个界面,在其中可以编辑 OpenXR Package Settings 资产文件中的配置,如图 7-6 所示,也可以直接编辑该资产文件,如图 7-7 所示。

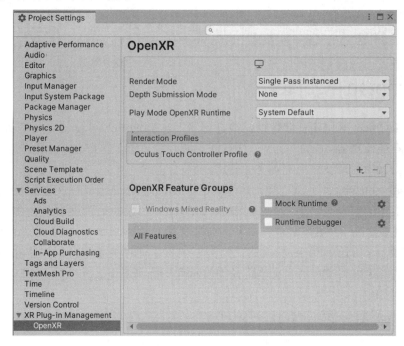

图 7-6　OpenXR 设置界面

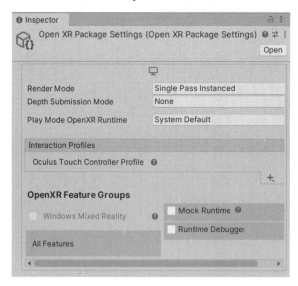

图 7-7　用于 OpenXR 设置的资产

那么如何注册自定义设置界面?需要用到 Settings Provider Attribute,将其用于一个返回值为 SettingsProvider 实例的静态方法。SettingsProvider 的构造函数的代码如下：

```
public SettingsProvider (string path, SettingsScope scopes,
IEnumerable<string> keywords);
```

参数 path 表示 Settings 窗口中设置的路径，使用 "/" 作为分隔符。参数 scopes 表示界面的作用范围类型，包含 User 与 Project 两种类型，前者表示在 Preferences 窗口中注册该设置界面，后者表示在 Project Settings 窗口中注册该设置界面。参数 keywords 则表示要与用户搜索内容进行比较的关键字列表，当用户在 Settings 窗口的搜索框中输入值时，SettingsProvider.HasSearchInterest()方法会尝试对这些关键字与此列表进行匹配，示例代码如下：

```
//第7章/ExampleSettingsProvider.cs

using UnityEngine;
using UnityEditor;
using System.Collections.Generic;

using UnityEngine;
using UnityEditor;
using System.Collections.Generic;

public class ExampleSettingsProvider : SettingsProvider
{
    [SettingsProvider]
    private static SettingsProvider CreateProvider()
    {
        return new ExampleSettingsProvider(
            "Project/Custom/Example Settings", SettingsScope.Project);
    }

    public ExampleSettingsProvider(string path,
        SettingsScope scopes, IEnumerable<string> keywords = null)
        : base(path, scopes, keywords) { }

    public override void OnGUI(string searchContext)
    {
        GUILayout.Label("Hello World.", EditorStyles.boldLabel);
        GUILayout.Button("Button");
    }
}
```

结果如图 7-8 所示。

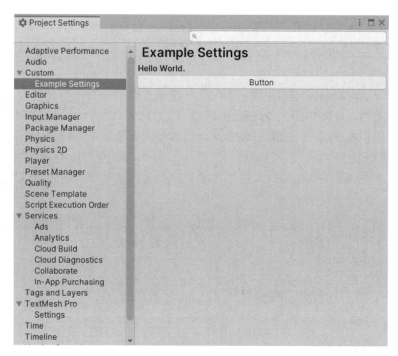

图 7-8 注册自定义设置界面

7.4 AssetModificationProcessor

AssetModificationProcessor 允许开发者在资产发生变化时执行指定的回调逻辑，该类包含的方法见表 7-4，使用时继承该类并实现对应的方法即可。

表 7-4 AssetModificationProcessor 中的回调方法

方　　法	作　　用
OnWillCreateAsset()	即将创建未导入的资产时调用此方法
OnWillDeleteAsset()	即将从磁盘中删除资产时会调用此方法
OnWillMoveAsset()	即将在磁盘上移动资产时会调用此方法
OnWillSaveAssets()	即将向磁盘写入序列化资产或场景文件时会调用此方法

当想要在创建 cs 脚本时自动地在脚本头部添加模板注释时可以使用 OnWillCreateAsset() 方法，首先判断如果创建的资产是 .cs 文件，则使用 File 类读取文件中的内容，然后在前面拼接模板注释的内容后再写入文件即可，代码如下：

```
//第 7 章/ScriptsHeader.cs

using System.IO;
using UnityEngine;
```

```csharp
//<summary>
//新创建的脚本自动添加模板注释(头部注释)
//</summary>
public class ScriptsHeader : UnityEditor.AssetModificationProcessor
{
    private const string author = "张寿昆";
    private const string email = "136512892@qq.com";
    private const string firstVersion = "1.0.0";

    //<summary>
    //资产创建时调用
    //</summary>
    //<param name="path">资产路径</param>
    public static void OnWillCreateAsset(string path)
    {
        path = path.Replace(".meta", "");
        if (!path.EndsWith(".cs")) return;
        string scriptPath = Application.dataPath
            .Replace("Assets", "") + path;
        string header = string.Format(
            "/*" 
            + "====================================================\r\n"
            + " * Unity版本：{0}\r\n"
            + " * 作者：{1}\r\n"
            + " * 邮箱：{2}\r\n"
            + " * 创建时间：{3}\r\n"
            + " * 当前版本：{4}\r\n"
            + " * 主要功能：\r\n"
            + " * 详细描述：\r\n"
            + " * 修改记录：\r\n"
            + " * "
            + "====================================================*/\r\n\r\n",
            Application.unityVersion, author, email,
            System.DateTime.Now.ToString(
                "yyyy-MM-dd HH:mm:ss"), firstVersion);
        File.WriteAllText(scriptPath, header
            + File.ReadAllText(scriptPath));
    }
}
```

有了 ScriptsHeader 脚本后再创建新的 cs 脚本文件时，其头部便会自动添加模板注释，如图 7-9 所示。

在日常开发工作中，可能会遇到不小心将一个文件夹拖到另一个文件夹里的情况，如果工程规模较大并且被移动的这个文件夹中包含脚本文件，则会开始漫长的脚本编译过程。

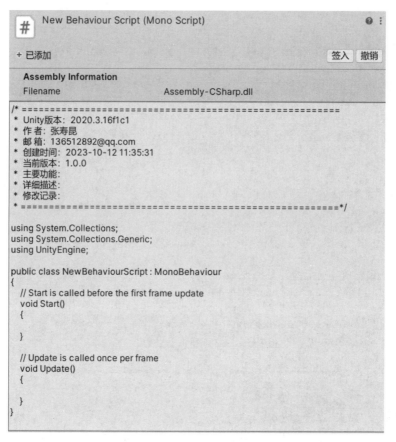

图 7-9 脚本自动添加模板注释

为了避免发生这种情况，可以使用 OnWillMoveAsset() 方法在资源即将被移动时判断文件夹中是否包含脚本文件，如果包含脚本文件，则弹出二次确认弹窗来确认移动操作。该方法的返回值为 AssetMoveResult 枚举类型，如果不允许移动资源，则需要使其返回值为 FailedMove 类型，代码如下：

```
//第7章/OnWillMoveAssetPrompt.cs
using UnityEditor;
using UnityEngine;

public class OnWillMoveAssetPrompt : UnityEditor.AssetModificationProcessor
{
    //<summary>
    //即将移动资产时调用
    //</summary>
    //<param name="sourcePath">资产的当前路径</param>
    //<param name="destinationPath">资产要移动到的路径</param>
    //<returns></returns>
```

```csharp
public static AssetMoveResult OnWillMoveAsset(
    string sourcePath, string destinationPath)
{
    //Debug.Log(string.Format("SourcePath: {0}  " +
    //"DestinationPath: {1}", sourcePath, destinationPath));
    if (AssetDatabase.IsValidFolder(sourcePath))
    {
        //在 sourcePath 文件夹中查找是否包含脚本文件
        if (AssetDatabase.FindAssets("t:Script",
            new string[] { sourcePath }).Length != 0)
        {
            //弹出确认弹窗
            if (!EditorUtility.DisplayDialog("提示", string.Format(
                "是否确认将文件夹{0}移动至{1}",
                sourcePath, destinationPath), "确认", "取消"))
                return AssetMoveResult.FailedMove;
        }
    }
    return AssetMoveResult.DidNotMove;
}
```

有了 OnWillMoveAssetPrompt 脚本后再移动包含脚本文件的文件夹便会弹出确认弹窗，如图 7-10 所示。

图 7-10　移动文件夹确认弹窗

7.5　AssetPostprocessor

AssetPostprocessor 允许开发者挂接到资产导入管线并在导入前后执行回调，该类中包含的回调方法见表 7-5，通过这些回调可以自定义处理资产的导入过程。

表 7-5　AssetPostprocessor 中的回调方法

方　　法	作　　用
OnPreprocessAsset()	在导入所有资产之前被调用
OnPostprocessAllAssets()	在导入所有资产之后被调用
OnPreprocessAudio()	在导入 AnimationClip 资产之前被调用

续表

方 法	作 用
OnPostprocessAudio()	在导入 AnimationClip 资产之后被调用
OnPreprocessTexture()	在导入 Texture 资产之前被调用
OnPostprocessTexture()	在导入 Texture 资产之后被调用
OnPreprocessAnimation()	在导入模型中的动画之前被调用
OnPostprocessAnimation()	在导入模型中的动画之后被调用
OnPreprocessModel()	在导入.fbx 等模型资产之前被调用
OnPostprocessModel()	在导入.fbx 等模型资产之后被调用
OnPreprocessMaterialDescription()	在从 Model Importer 导入材质之前被调用
OnPostprocessMaterial()	在导入 Material 资产之后被调用
OnPreprocessSpeedTree()	在导入 SpeedTree 资产之前被调用
OnPostprocessSpeedTree()	在导入 SpeedTree 资产之后被调用
OnPostprocessAssetbundleNameChanged()	在 AssetBundle 名称发生变更时被调用
OnPostprocessCubemap()	在导入 Cubemap 资产之后被调用
OnPostprocessPrefab()	在导入 Prefab 资产之后被调用
OnPostprocessSprites()	在导入 Sprite 资产之后被调用
OnPostprocessGameObjectWithAnimatedUserProperties()	在导入自定义属性的动画曲线之后被调用
OnPostprocessGameObjectWithUserProperties()	为每个在导入文件中至少附加了一个用户属性的游戏对象调用此函数
OnPostprocessMeshHierarchy()	当变换层级视图已完成导入时调用此函数

资产导入时的路径通过变量 assetPath 获取，对应的资产导入器通过变量 assetImporter 获取，可以通过类型转换将其转换为相应的类型。例如在导入模型资产时，将其转换为 ModelImporter 类型，在导入音频资产时，将其转换为 AudioImporter 类型，这样就可以对相应资产的导入设置进行调整。

在 Texture 资产的导入设置中，通过减小 Texture 的尺寸并适当地进行压缩，可以减少资产占用的大小，所以在资产管理及性能优化的工作中，通常会将 Texture 降到合适的尺寸并进行适当压缩。

除此之外，也会考虑关闭 Read/Write Enabled 的设置，一般情况下不需要可读写设置，如果启用，则在 CPU 和 GPU 可寻址内存中都会创建副本，Texture 将会占用双倍内存。Minimap 也会占用额外的空间，对于在屏幕上大小保持不变的 Texture，它是非必要的，例如 UI，因此在 UI 界面搭建的过程中一般会禁用 Minimap，并且在导入 UI 切图资产后，需要将 Texture Type 设置为 Sprite 类型。

有了 AssetPostprocessor 便可以通过 OnPreprocessTexture()方法在 Texture 资产导入时自动完成以上设置，Texture Import Settings 中的属性如图 7-11 所示。

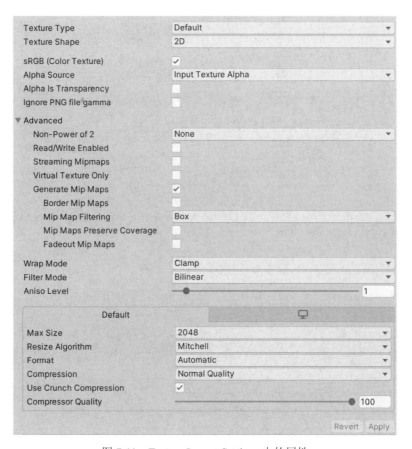

图 7-11 Texture Import Settings 中的属性

根据这些属性编写配置脚本，将导入配置存储在 ScritableObject 类型资产中，避免在有变更需求时去修改代码来完成配置，代码如下：

```
//第7章/TexturePreprocessSettings.cs

using System.IO;
using UnityEngine;
using UnityEditor;

[CreateAssetMenu]
public class TexturePreprocessSettings : ScriptableObject
{
    [SerializeField] private TextureImporterType textureType
        = TextureImporterType.Default;
    [SerializeField] private TextureImporterShape textureShape
        = TextureImporterShape.Texture2D;
    [SerializeField] private bool sRGBTexture = true;
    [SerializeField] private TextureImporterAlphaSource alphaSource
```

```csharp
        = TextureImporterAlphaSource.FromInput;
    [SerializeField] private bool alphaIsTransparency;
    [SerializeField] private bool ignorePNGFileGamma;

    [Header("Advanced")]
    [SerializeField] private TextureImporterNPOTScale nonPowerOf2
        = TextureImporterNPOTScale.ToNearest;
    [SerializeField] private bool readWriteEnabled;
    [SerializeField] private bool streamingMipmaps;
    [SerializeField] private bool vitrualTextureOnly;
    [SerializeField] private bool generateMipMaps = true;
    [SerializeField] private bool borderMipMaps;
    [SerializeField] private TextureImporterMipFilter mipmapFilter
        = TextureImporterMipFilter.BoxFilter;
    [SerializeField] private bool mipMapsPreserveCoverage;
    [SerializeField] private bool fadeoutMipMaps;

    [SerializeField] private TextureWrapMode wrapMode
        = TextureWrapMode.Repeat;
    [SerializeField] private FilterMode filterMode
        = FilterMode.Bilinear;
    [SerializeField, Range(0, 16)] private int anisoLevel = 1;

    [SerializeField] private int maxSize = 2048;
    [SerializeField] private TextureImporterFormat format
        = TextureImporterFormat.Automatic;
    [SerializeField] private TextureImporterCompression compression
        = TextureImporterCompression.Compressed;
    [SerializeField] private bool useCrunchCompression;

    private static TexturePreprocessSettings m_Settings;
    private static TexturePreprocessSettings Settings
    {
        get
        {
            if (m_Settings == null)
            {
                var path = "Assets/Settings/" +
                    "Texture Preprocess Settings.asset";
                m_Settings = AssetDatabase
                    .LoadAssetAtPath<TexturePreprocessSettings>(path);
                if (m_Settings == null)
                {
                    m_Settings = CreateInstance<TexturePreprocessSettings>();
                    var directory = Application.dataPath + "/Settings";
                    if (!Directory.Exists(directory))
                        Directory.CreateDirectory(
```

```csharp
                    Application.dataPath + "/Settings");
                AssetDatabase.CreateAsset(m_Settings, path);
                AssetDatabase.Refresh();
            }
        }
        return m_Settings;
    }
}

public static TextureImporterType TextureType {
    get { return Settings.textureType; } }
public static TextureImporterShape TextureShape {
    get { return Settings.textureShape; } }
public static bool SRGBTexture {
    get { return Settings.sRGBTexture; } }
public static TextureImporterAlphaSource AlphaSource {
    get { return Settings.alphaSource; } }
public static bool AlphaIsTransparency {
    get { return Settings.alphaIsTransparency; } }
public static bool IgnorePNGFileGamma {
    get { return Settings.ignorePNGFileGamma; } }

public static TextureImporterNPOTScale NonPowerOf2 {
    get { return Settings.nonPowerOf2; } }
public static bool ReadWriteEnabled {
    get { return Settings.readWriteEnabled; } }
public static bool StreamingMipmaps {
    get { return Settings.streamingMipmaps; } }
public static bool VitrualTextureOnly {
    get { return Settings.vitrualTextureOnly; } }
public static bool GenerateMipMaps {
    get { return Settings.generateMipMaps; } }
public static bool BorderMipMaps {
    get { return Settings.borderMipMaps; } }
public static TextureImporterMipFilter MipmapFilter {
    get { return Settings.mipmapFilter; } }
public static bool MipMapsPreserveCoverage {
    get { return Settings.mipMapsPreserveCoverage; } }
public static bool FadeoutMipMaps {
    get { return Settings.fadeoutMipMaps; } }

public static TextureWrapMode WrapMode {
    get { return Settings.wrapMode; } }
public static FilterMode FilterMode {
    get { return Settings.filterMode; } }
public static int AnisoLevel {
    get { return Settings.anisoLevel; } }
```

```csharp
    public static int MaxSize {
        get { return Settings.maxSize; } }
    public static TextureImporterFormat Format {
        get { return Settings.format; } }
    public static TextureImporterCompression Compression {
        get { return Settings.compression; } }
    public static bool UseCrunchCompression {
        get { return Settings.useCrunchCompression; } }
}
```

有了该脚本后，选择 Assets→Create→Texture Preprocess Settings 命令创建该类型的资产并存放到指定目录，如图 7-12 所示。

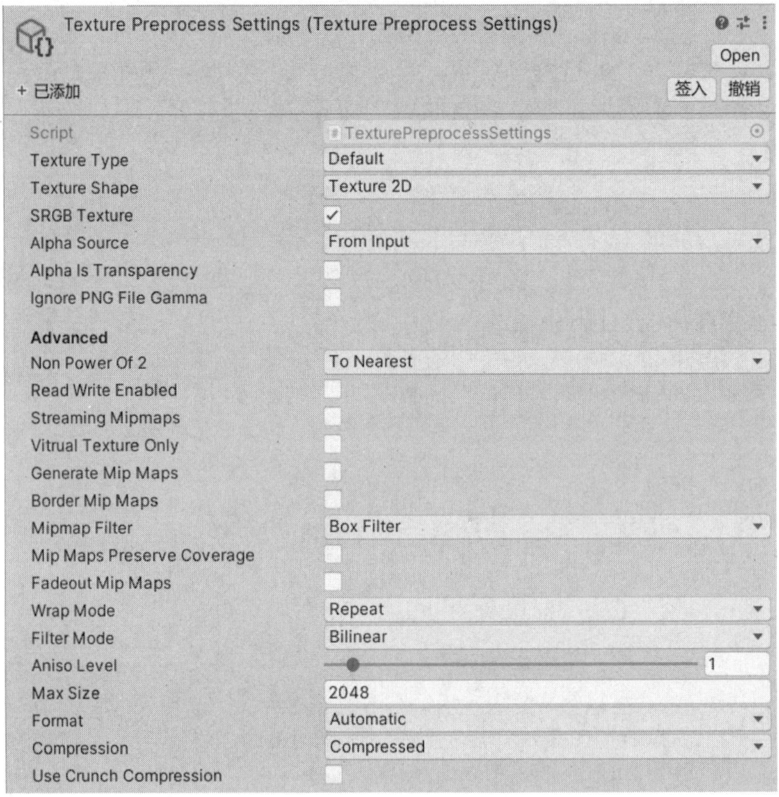

图 7-12　Texture Preprocess Settings

在 OnPreprocessTexture()方法中可以根据这个配置文件对导入的 Texture 资产进行各属性的设置，代码如下：

```csharp
//第 7 章/TexturePostprocessor.cs

using UnityEngine;
```

```csharp
using UnityEditor;

public class TexturePostprocessor : AssetPostprocessor
{
    private void OnPreprocessTexture()
    {
        TextureImporter importer = assetImporter as TextureImporter;
        if (importer == null) return;

        importer.textureShape = TexturePreprocessSettings.TextureShape;
        importer.sRGBTexture = TexturePreprocessSettings.SRGBTexture;
        importer.alphaSource = TexturePreprocessSettings.AlphaSource;
        importer.alphaIsTransparency =
            TexturePreprocessSettings.AlphaIsTransparency;
        importer.ignorePngGamma =
            TexturePreprocessSettings.IgnorePNGFileGamma;
        importer.npotScale = TexturePreprocessSettings.NonPowerOf2;
        importer.isReadable = TexturePreprocessSettings.ReadWriteEnabled;
        importer.streamingMipmaps =
            TexturePreprocessSettings.StreamingMipmaps;
        importer.vtOnly = TexturePreprocessSettings.VitrualTextureOnly;
        importer.mipmapEnabled = TexturePreprocessSettings.GenerateMipMaps;
        importer.borderMipmap = TexturePreprocessSettings.BorderMipMaps;
        importer.mipmapFilter = TexturePreprocessSettings.MipmapFilter;
        importer.mipMapsPreserveCoverage =
            TexturePreprocessSettings.MipMapsPreserveCoverage;
        importer.fadeout = TexturePreprocessSettings.FadeoutMipMaps;
        importer.wrapMode = TexturePreprocessSettings.WrapMode;
        importer.filterMode = TexturePreprocessSettings.FilterMode;
        importer.anisoLevel = TexturePreprocessSettings.AnisoLevel;
        importer.maxTextureSize = TexturePreprocessSettings.MaxSize;
        importer.textureCompression =
            TexturePreprocessSettings.Compression;
        importer.crunchedCompression =
            TexturePreprocessSettings.UseCrunchCompression;
        importer.textureType = TexturePreprocessSettings.TextureType;
    }
}
```

7.6 BuildPipeline

BuildPipeline 类是 Unity 中用于构建 AssetBundle 资源包和构建应用程序包体的工具类。它提供了一系列静态方法，开发者可以通过该类在编辑器中使构建和打包流程自动化，以便在不同平台上部署。这个类还提供了一些额外的功能，例如自定义构建的设置和自动化脚本的执行。

总体来讲，BuildPipeline 类是 Unity 中非常重要的工具，可以帮助开发者简化打包和发布的流程，本节通过两个示例展示该类中相关方法的应用。

7.6.1 AssetBundle 构建工具

用于构建 AssetBundle 资源包的方法的代码如下，参数 outputPath 表示 Asset Bundle 的输出路径，assetBundleOptions 表示构建 Asset Bundle 的选项，targetPaltform 表示 Asset Bundle 的目标构建平台。

```
public static AssetBundleManifest BuildAssetBundles (string outputPath,
BuildAssetBundleOptions assetBundleOptions, BuildTarget targetPlatform);
```

本节通过该方法实现一个构建 AssetBundle 资源包的工具，如图 7-13 所示。

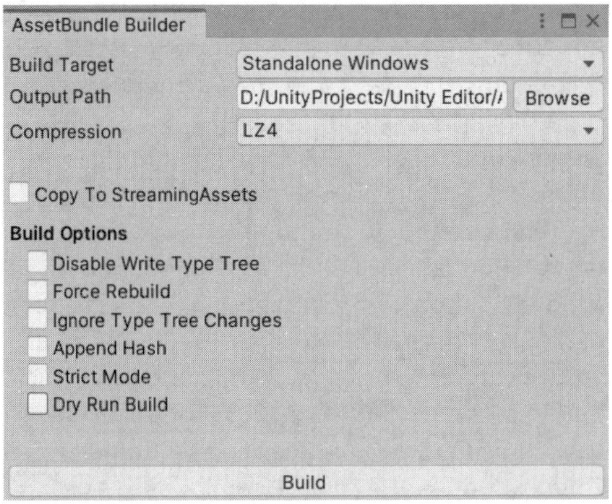

图 7-13　AssetBundle 资源包的工具

首先通过 Enum Popup 控件提供目标平台、压缩方式的选项，其中压缩方式有 3 种类型，分别是使用 LZ4 算法压缩、使用 LZMA 算法压缩及不压缩，它们对应于 BuildAssetBundleOptions 中的 ChunkBasedCompression、None、UncompressedAssetBundle 枚举值，其中 LZ4 是最常用的压缩算法，它的压缩率没有 LZMA 算法的压缩率高，但是在加载时它可以加载指定的资源包而不需要解压全部，因此加载速度较快，而且比不压缩的资源包包体要小。

然后使用 Toggle 控件提供是否将资源包复制至 StreamingAssets 文件夹等选项，最终提供一个按钮控件，当单击该按钮时，调用 BuildAssetBundles()方法，开始构建资源包，代码如下：

```
//第 7 章/AssetBundleBuilder.cs

using System;
using System.IO;
using System.Text;
```

```csharp
using System.Collections.Generic;
using System.Runtime.Serialization.Formatters.Binary;

using UnityEditor;
using UnityEngine;

public class AssetBundleBuilder : EditorWindow
{
    [MenuItem("Example/Resource/AssetBundle Builder")]
    public static void Open()
    {
        GetWindow<AssetBundleBuilder>("AssetBundle Builder").Show();
    }

    [SerializeField] private BuildTabData data;
    private Vector2 scroll;

    private void OnEnable()
    {
        //数据文件路径
        string dataPath = Path.GetFullPath(".")
            .Replace("\\", "/") + "/Library/AssetBundleBuild.dat";
        //判断数据文件是否存在
        if (File.Exists(dataPath))
        {
            //打开文件
            using (FileStream fs = File.Open(dataPath, FileMode.Open))
            {
                //反序列化
                BinaryFormatter bf = new BinaryFormatter();
                var deserialize = bf.Deserialize(fs);
                if (deserialize != null)
                    data = deserialize as BuildTabData;
                if (data == null)
                {
                    File.Delete(dataPath);
                    data = new BuildTabData()
                    {
                        buildTarget = BuildTarget.StandaloneWindows,
                        outputPath = Application.streamingAssetsPath,
                        compressionType = BuildTabData.CompressionType.LZ4
                    };
                }
            }
        }
        else
        {
```

```csharp
            data = new BuildTabData()
            {
                buildTarget = BuildTarget.StandaloneWindows,
                outputPath = Application.streamingAssetsPath,
                compressionType = BuildTabData.CompressionType.LZ4
            };
        }
    }

    private void OnDisable()
    {
        //数据文件路径
        string dataPath = Path.GetFullPath(".")
            .Replace("\\", "/") + "/Library/AssetBundleBuild.dat";
        //对写入的数据文件进行保存
        using (FileStream fs = File.Create(dataPath))
        {
            //序列化
            BinaryFormatter bf = new BinaryFormatter();
            bf.Serialize(fs, data);
        }
    }

    private void OnGUI()
    {
        EditorGUILayout.Space();
        scroll = EditorGUILayout.BeginScrollView(scroll);
        //目标平台
        data.buildTarget = (BuildTarget)EditorGUILayout
            .EnumPopup("Build Target", data.buildTarget);
        GUILayout.BeginHorizontal();
        //输出路径
        data.outputPath = EditorGUILayout
            .TextField("Output Path", data.outputPath);
        //浏览按钮
        if (GUILayout.Button("Browse", GUILayout.Width(60f)))
        {
            //选择输出路径
            string selectedPath = EditorUtility.OpenFolderPanel(
                "AssetsBundle构建路径", Application.dataPath, string.Empty);
            //判断没有Cancel(路径不为空),更新输出路径
            if (!string.IsNullOrEmpty(selectedPath))
                data.outputPath = selectedPath;
        }
        GUILayout.EndHorizontal();
        //压缩方式
        data.compressionType = (BuildTabData.CompressionType)
```

```csharp
            EditorGUILayout.EnumPopup("Compression", data.compressionType);
        GUILayout.Space(20f);
        //是否复制至 Assets/StreamingAssets
        data.copy2StreamingAssets = GUILayout.Toggle(
            data.copy2StreamingAssets, GUIContents.copy2StreamingAssets);
        EditorGUILayout.Space();
        //Options
        GUILayout.Label("Build Options", "BoldLabel");
        GUILayout.BeginHorizontal();
        GUILayout.Space(15f);
        GUILayout.BeginVertical();
        data.disableWriteTypeTree = GUILayout.Toggle(
            data.disableWriteTypeTree,
            GUIContents.disableWriteTypeTree);
        data.forceRebuildAssetBundle = GUILayout.Toggle(
            data.forceRebuildAssetBundle,
            GUIContents.forceRebuildAssetBundle);
        data.ignoreTypeTreeChanges = GUILayout.Toggle(
            data.ignoreTypeTreeChanges,
            GUIContents.ignoreTypeTreeChanges);
        data.appendHashToAssetBundleName = GUILayout.Toggle(
            data.appendHashToAssetBundleName,
            GUIContents.appendHash);
        data.strictMode = GUILayout.Toggle(
            data.strictMode,
            GUIContents.strictMode);
        data.dryRunBuild = GUILayout.Toggle(
            data.dryRunBuild,
            GUIContents.dryRunBuild);
        GUILayout.EndVertical();
        GUILayout.EndHorizontal();
        GUILayout.FlexibleSpace();
        //构建按钮
        if (GUILayout.Button("Build"))
        {
            //提醒
            if (EditorUtility.DisplayDialog("提醒",
                "构建 AssetsBundle 将花费一定时间，是否确定开始？", "确定", "取消"))
                BuildAssetBundle();
        }
        EditorGUILayout.EndScrollView();
}

//将一个文件夹中的内容复制到另一个文件夹
internal void CopyDirectory(string sourceDir, string destDir)
{
    //如果目标文件夹不存在，则创建文件夹
```

```csharp
        if (!Directory.Exists(destDir))
            Directory.CreateDirectory(destDir);
        //获取源文件夹中的所有文件夹
        string[] directories = Directory.GetDirectories(
            sourceDir, "*", SearchOption.AllDirectories);
        //遍历源文件夹中的所有文件夹
        for (int i = 0; i < directories.Length; i++)
        {
            //如果目标文件夹中不存在与源文件夹中对应的文件夹，则创建
            string targetDir = directories[i].Replace(sourceDir, destDir);
            if (!Directory.Exists(targetDir))
                Directory.CreateDirectory(targetDir);
        }
        //获取源文件夹中的所有文件
        string[] files = Directory.GetFiles(sourceDir,
            "*.*", SearchOption.AllDirectories);
        //遍历源文件夹中的所有文件
        for (int i = 0; i < files.Length; i++)
        {
            string filePath = files[i];
            string fileDirName = Path
                .GetDirectoryName(filePath).Replace("\\", "/");
            string fileName = Path.GetFileName(filePath);
            //路径拼接
            string newFilePath = Path.Combine(
                fileDirName.Replace(sourceDir, destDir), fileName);
            //复制文件
            File.Copy(filePath, newFilePath, true);
        }
    }

    private void BuildAssetBundle()
    {
        //检查路径是否存在
        if (!Directory.Exists(data.outputPath))
        {
            Debug.Log(string.Format("路径不存在，进行创建 {0}",
                data.outputPath));
            Directory.CreateDirectory(data.outputPath);
        }
        //Options
        BuildAssetBundleOptions options
            = BuildAssetBundleOptions.ChunkBasedCompression;
        switch (data.compressionType)
        {
            case BuildTabData.CompressionType.Uncompressed:
                options = BuildAssetBundleOptions.UncompressedAssetBundle;
```

```csharp
            break;
        case BuildTabData.CompressionType.LZMA:
            options = BuildAssetBundleOptions.None;
            break;
        case BuildTabData.CompressionType.LZ4:
            options = BuildAssetBundleOptions.ChunkBasedCompression;
            break;
    }
    if (data.disableWriteTypeTree)
        options |= BuildAssetBundleOptions.DisableWriteTypeTree;
    if (data.forceRebuildAssetBundle)
        options |= BuildAssetBundleOptions.ForceRebuildAssetBundle;
    if (data.ignoreTypeTreeChanges)
        options |= BuildAssetBundleOptions.IgnoreTypeTreeChanges;
    if (data.appendHashToAssetBundleName)
        options |= BuildAssetBundleOptions.AppendHashToAssetBundleName;
    if (data.strictMode)
        options |= BuildAssetBundleOptions.StrictMode;
    if (data.dryRunBuild)
        options |= BuildAssetBundleOptions.DryRunBuild;

    //开始构建
    AssetBundleManifest manifest = BuildPipeline
        .BuildAssetBundles(data.outputPath, options, data.buildTarget);
    //Map
    List<AssetInfo> map = new List<AssetInfo>();
    string[] assetBundleNames = manifest.GetAllAssetBundles();
    for (int i = 0; i < assetBundleNames.Length; i++)
    {
        string[] paths = AssetDatabase
            .GetAssetPathsFromAssetBundle(assetBundleNames[i]);
        for (int j = 0; j < paths.Length; j++)
        {
            map.Add(new AssetInfo(paths[j], assetBundleNames[i]));
        }
    }
    string json = JsonUtility.ToJson(new AssetsInfo(map));
    byte[] buffer = Encoding.Default.GetBytes(json);
    string mapPath = Path.Combine(data.outputPath, "map.dat");
    using (FileStream fs = File.Create(mapPath))
    {
        BinaryFormatter bf = new BinaryFormatter();
        bf.Serialize(fs, buffer);
    }
    //复制至Assets/StreamingAssets
    if (data.copy2StreamingAssets)
    {
```

```csharp
            //如果输出路径本身已经是StreamingAssets路径，则不处理
            if (!data.outputPath.StartsWith(
                Application.streamingAssetsPath))
            {
                CopyDirectory(data.outputPath,
                    Application.streamingAssetsPath);
            }
        }
        AssetDatabase.Refresh();
}

//数据类
[Serializable]
internal class BuildTabData
{
    internal enum CompressionType { Uncompressed, LZMA, LZ4 }

    //输出路径
    internal string outputPath;
    //压缩方式
    internal CompressionType compressionType;
    //目标平台
    internal BuildTarget buildTarget;
    //是否复制至Assets/StreamingAssets
    internal bool copy2StreamingAssets;
    //Options
    internal bool disableWriteTypeTree;
    internal bool forceRebuildAssetBundle;
    internal bool ignoreTypeTreeChanges;
    internal bool appendHashToAssetBundleName;
    internal bool strictMode;
    internal bool dryRunBuild;
}
//<summary>
//Asset 资产信息
//</summary>
[Serializable]
public class AssetInfo
{
    //<summary>
    //资源路径
    //</summary>
    public string path;

    //<summary>
    //AssetBundle 包名称
    //</summary>
```

```csharp
        public string abName;

        public AssetInfo(string path, string abName)
        {
            this.path = path;
            this.abName = abName;
        }
    }
    [Serializable]
    public class AssetsInfo
    {
        public List<AssetInfo> list = new List<AssetInfo>();

        public AssetsInfo(List<AssetInfo> list)
        {
            this.list = list;
        }
    }

    private class GUIContents
    {
        public static GUIContent copy2StreamingAssets = new GUIContent(
            "Copy To StreamingAssets",
            "Copy asset bundle to Assets/StreamingAssets after " +
            "build completed for use in simulation mode.");
        public static GUIContent disableWriteTypeTree = new GUIContent(
            "Disable Write Type Tree",
            "Do not include type information within the AssetBundle.");
        public static GUIContent forceRebuildAssetBundle = new GUIContent(
            "Force Rebuild",
            "Force rebuild the assetBundles.");
        public static GUIContent ignoreTypeTreeChanges = new GUIContent(
            "Ignore Type Tree Changes",
            "Ignore the type tree changes when " +
            "doing the incremental build check.");
        public static GUIContent appendHash = new GUIContent(
            "Append Hash",
            "Append the hash to to asset bundle name.");
        public static GUIContent strictMode = new GUIContent(
            "Strict Mode",
            " Do not allow the build to succeed if " +
            "any errors are reporting during it.");
        public static GUIContent dryRunBuild = new GUIContent(
            "Dry Run Build",
            "Do a dry run build.");
    }
}
```

7.6.2 应用程序批量构建工具

BuildPipeline 中用于构建应用程序的方法的代码如下,参数 levels 表示构建的应用程序包含的场景,locationPathName 表示构建的应用程序的输出路径,target 表示构建的目标平台,options 表示构建选项。

```
public static Build.Reporting.BuildReport BuildPlayer (string[] levels,
string locationPathName, BuildTarget target, BuildOptions options);
    public static Build.Reporting.BuildReport BuildPlayer
(EditorBuildSettingsScene[] levels, string locationPathName, BuildTarget target,
BuildOptions options);
```

通常情况下,如果项目对应的是 PC、Android 或 iOS 端,则都不会有批量构建应用程序这样的需求,但如果项目对应的是 WebGL 端,则可能会遇到这样的需求:将不同的场景打包成不同的应用程序,入口是在前端 Web 页面中布局的,单击会打开相应的 WebGL 程序链接。

在这种情况下,项目在构建时就需要构建多个包体,依次手动打包比较烦琐,而且需要等待很长时间。本节通过创建一个配置文件,在其中配置构建不同包体时的数据,例如场景、名称、输出路径等,通过遍历这些数据,调用 BuildPipeline 中的 BuildPlayer()方法,就可以实现批量构建应用程序。

首先定义每项构建任务的数据类,该类需要支持序列化,代码如下:

```
//第7章/BuildTask.cs

using System;
using UnityEditor;
using System.Collections.Generic;

[Serializable]
public sealed class BuildTask
{
    //<summary>
    //名称
    //</summary>
    public string ProductName;
    //<summary>
    //目标平台
    //</summary>
    public BuildTarget BuildTarget;
    //<summary>
    //打包路径
    //</summary>
    public string BuildPath;
    //<summary>
```

```
    //打包场景
    //</summary>
    public List<SceneAsset> SceneAssets = new List<SceneAsset>(0);
}
```

使用 ScriptableObject 作为配置的数据容器，代码如下：

```
//第7章/BuildTask.cs

using UnityEngine;
using System.Collections.Generic;

[CreateAssetMenu(fileName = "New Build Profile", menuName = "Build Profile")]
public sealed class BuildProfile : ScriptableObject
{
    //<summary>
    //打包任务列表
    //</summary>
    public List<BuildTask> BuildTasks = new List<BuildTask>(0);
}
```

有了配置文件后，配置构建列表，批量构建要做的就是遍历该列表并依次调用 BuildPipeline 中的 BuildPlayer() 方法。创建配置文件类的自定义编辑器，编写打包功能，以及添加、移除打包任务项等菜单，代码如下：

```
//第7章/BuildProfileEditor.cs

using System.IO;
using System.Text;
using System.Collections.Generic;

using UnityEngine;
using UnityEditor;

[CustomEditor(typeof(BuildProfile))]
public sealed class BuildProfileEditor : Editor
{
    private readonly Dictionary<BuildTask, bool> foldoutMap
        = new Dictionary<BuildTask, bool>();
    private Vector2 scroll = Vector2.zero;
    private BuildProfile profile;

    private void OnEnable()
    {
        profile = target as BuildProfile;
    }
    public override void OnInspectorGUI()
    {
```

```csharp
        OnTopGUI();
        OnBodyGUI();

        if (GUI.changed)
        {
            serializedObject.ApplyModifiedProperties();
            EditorUtility.SetDirty(profile);
        }
    }

    private void OnTopGUI()
    {
        GUILayout.BeginHorizontal();
        {
            if (GUILayout.Button("新建", EditorStyles.miniButtonLeft))
            {
                Undo.RecordObject(profile, "Create");
                var task = new BuildTask()
                {
                    ProductName = "Product Name",
                    BuildTarget = BuildTarget.StandaloneWindows64,
                    BuildPath = Directory.GetParent(
                        Application.dataPath).FullName
                };
                profile.BuildTasks.Add(task);
            }
            if (GUILayout.Button("展开", EditorStyles.miniButtonMid))
            {
                for (int i = 0; i < profile.BuildTasks.Count; i++)
                {
                    foldoutMap[profile.BuildTasks[i]] = true;
                }
            }
            if (GUILayout.Button("收缩", EditorStyles.miniButtonMid))
            {
                for (int i = 0; i < profile.BuildTasks.Count; i++)
                {
                    foldoutMap[profile.BuildTasks[i]] = false;
                }
            }
            GUI.color = Color.yellow;
            if (GUILayout.Button("清空", EditorStyles.miniButtonMid))
            {
                Undo.RecordObject(profile, "Clear");
                if (EditorUtility.DisplayDialog("提醒",
                    "是否确定清空列表?", "确定", "取消"))
                {
```

```csharp
                profile.BuildTasks.Clear();
            }
        }
        GUI.color = Color.cyan;
        if (GUILayout.Button("打包", EditorStyles.miniButtonRight))
        {
            if (EditorUtility.DisplayDialog("提醒",
                "打包需要耗费一定时间,是否确定开始?", "确定", "取消"))
            {
                StringBuilder sb = new StringBuilder();
                sb.Append("打包报告:\r\n");
                for (int i = 0; i < profile.BuildTasks.Count; i++)
                {
                    EditorUtility.DisplayProgressBar("Build",
                        "Building...", (float)(i + 1) /
                            profile.BuildTasks.Count);
                    var task = profile.BuildTasks[i];
                    List<EditorBuildSettingsScene> buildScenes
                        = new List<EditorBuildSettingsScene>();
                    for (int j = 0; j < task.SceneAssets.Count; j++)
                    {
                        var scenePath = AssetDatabase
                            .GetAssetPath(task.SceneAssets[j]);
                        if (!string.IsNullOrEmpty(scenePath))
                        {
                            buildScenes.Add(
                                new EditorBuildSettingsScene(
                                    scenePath, true));
                        }
                    }
                    string locationPathName = string.Format("{0}/{1}",
                        task.BuildPath, task.ProductName);
                    var report = BuildPipeline.BuildPlayer(
                        buildScenes.ToArray(), locationPathName,
                        task.BuildTarget, BuildOptions.None);
                    sb.Append(string.Format("[{0}] 打包结果: {1}\r\n",
                        task.ProductName, report.summary.result));
                }
                EditorUtility.ClearProgressBar();
                Debug.Log(sb.ToString());
            }
            return;
        }
        GUI.color = Color.white;
    }
    GUILayout.EndHorizontal();
}
```

```csharp
private void OnBodyGUI()
{
    scroll = GUILayout.BeginScrollView(scroll);
    {
        for (int i = 0; i < profile.BuildTasks.Count; i++)
        {
            var task = profile.BuildTasks[i];
            if (!foldoutMap.ContainsKey(task))
                foldoutMap.Add(task, true);
            GUILayout.BeginHorizontal("Badge");
            GUILayout.Space(12);
            foldoutMap[task] = EditorGUILayout
                .Foldout(foldoutMap[task], $"{task.ProductName}", true);
            GUILayout.Label(string.Empty);
            if (GUILayout.Button(EditorGUIUtility
                .IconContent("TreeEditor.Trash"),
                "IconButton", GUILayout.Width(20)))
            {
                Undo.RecordObject(profile, "Delete Task");
                foldoutMap.Remove(task);
                profile.BuildTasks.Remove(task);
                break;
            }
            GUILayout.EndHorizontal();
            if (foldoutMap[task])
            {
                GUILayout.BeginVertical("Box");
                GUILayout.BeginHorizontal();
                GUILayout.Label("打包场景: ", GUILayout.Width(70));
                if (GUILayout.Button(EditorGUIUtility
                    .IconContent("Toolbar Plus More"),
                    GUILayout.Width(28)))
                {
                    task.SceneAssets.Add(null);
                }
                GUILayout.EndHorizontal();
                if (task.SceneAssets.Count > 0)
                {
                    GUILayout.BeginHorizontal();
                    GUILayout.Space(75);
                    GUILayout.BeginVertical("Badge");
                    for (int j = 0; j < task.SceneAssets.Count; j++)
                    {
                        var sceneAsset = task.SceneAssets[j];
                        GUILayout.BeginHorizontal();
                        GUILayout.Label($"{j + 1}.", GUILayout.Width(20));
                        task.SceneAssets[j] = EditorGUILayout
```

```csharp
        .ObjectField(sceneAsset, typeof(SceneAsset),
            false) as SceneAsset;
    if (GUILayout.Button("↑",
        EditorStyles.miniButtonLeft,
        GUILayout.Width(20)))
    {
        if (j > 0)
        {
            Undo.RecordObject(profile,
                "Move Up Scene Assets");
            var temp = task.SceneAssets[j - 1];
            task.SceneAssets[j - 1] = sceneAsset;
            task.SceneAssets[j] = temp;
        }
    }
    if (GUILayout.Button("↓",
        EditorStyles.miniButtonMid,
        GUILayout.Width(20)))
    {
        if (j < task.SceneAssets.Count - 1)
        {
            Undo.RecordObject(profile,
                "Move Down Scene Assets");
            var temp = task.SceneAssets[j + 1];
            task.SceneAssets[j + 1] = sceneAsset;
            task.SceneAssets[j] = temp;
        }
    }
    if (GUILayout.Button(EditorGUIUtility
        .IconContent("Toolbar Plus"),
        EditorStyles.miniButtonMid,
        GUILayout.Width(20)))
    {
        Undo.RecordObject(profile,
            "Add Scene Assets");
        task.SceneAssets.Insert(j + 1, null);
        break;
    }
    if (GUILayout.Button(EditorGUIUtility
        .IconContent("Toolbar Minus"),
        EditorStyles.miniButtonRight,
        GUILayout.Width(20)))
    {
        Undo.RecordObject(profile,
            "Delete Scene Assets");
        task.SceneAssets.RemoveAt(j);
        break;
```

```
                    }
                    GUILayout.EndHorizontal();
                }
                GUILayout.EndVertical();
                GUILayout.EndHorizontal();
            }
            GUILayout.BeginHorizontal();
            GUILayout.Label("产品名称：", GUILayout.Width(70));
            var newPN = GUILayout.TextField(task.ProductName);
            if (task.ProductName != newPN)
            {
                Undo.RecordObject(profile, "Product Name");
                task.ProductName = newPN;
            }
            GUILayout.EndHorizontal();
            GUILayout.BeginHorizontal();
            GUILayout.Label("打包平台：", GUILayout.Width(70));
            var newBT = (BuildTarget)EditorGUILayout
                .EnumPopup(task.BuildTarget);
            if (task.BuildTarget != newBT)
            {
                Undo.RecordObject(profile, "Build Target");
                task.BuildTarget = newBT;
            }
            GUILayout.EndHorizontal();
            GUILayout.BeginHorizontal();
            GUILayout.Label("打包路径：", GUILayout.Width(70));
            GUILayout.TextField(task.BuildPath);
            if (GUILayout.Button("Browse", GUILayout.Width(60)))
            {
                task.BuildPath = EditorUtility
                    .SaveFolderPanel("Build Path",
                        task.BuildPath, "");
            }
            GUILayout.EndHorizontal();
            GUILayout.EndVertical();
        }
    }
}
        GUILayout.EndScrollView();
    }
}
```

最终效果如图 7-14 所示。

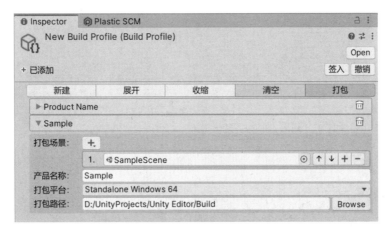

图 7-14　应用程序批量构建工具

图 书 推 荐

书 名	作 者
深度探索 Vue.js——原理剖析与实战应用	张云鹏
前端三剑客——HTML5+CSS3+JavaScript 从入门到实战	贾志杰
剑指大前端全栈工程师	贾志杰、史广、赵东彦
Flink 原理深入与编程实战——Scala+Java（微课视频版）	辛立伟
Spark 原理深入与编程实战（微课视频版）	辛立伟、张帆、张会娟
PySpark 原理深入与编程实战（微课视频版）	辛立伟、辛雨桐
HarmonyOS 移动应用开发（ArkTS 版）	刘安战、余雨萍、陈争艳 等
HarmonyOS 应用开发实战（JavaScript 版）	徐礼文
HarmonyOS 原子化服务卡片原理与实战	李洋
鸿蒙操作系统开发入门经典	徐礼文
鸿蒙应用程序开发	董昱
鸿蒙操作系统应用开发实践	陈美汝、郑森文、武延军、吴敬征
HarmonyOS 移动应用开发	刘安战、余雨萍、李勇军 等
HarmonyOS App 开发从 0 到 1	张诏添、李凯杰
JavaScript 修炼之路	张云鹏、戚爱斌
JavaScript 基础语法详解	张旭乾
华为方舟编译器之美——基于开源代码的架构分析与实现	史宁宁
Android Runtime 源码解析	史宁宁
数字 IC 设计入门（微课视频版）	白栎旸
数字电路设计与验证快速入门——Verilog+SystemVerilog	马骁
鲲鹏架构入门与实战	张磊
鲲鹏开发套件应用快速入门	张磊
华为 HCIA 路由与交换技术实战	江礼教
华为 HCIP 路由与交换技术实战	江礼教
openEuler 操作系统管理入门	陈争艳、刘安战、贾玉祥 等
5G 核心网原理与实践	易飞、何宇、刘子琦
恶意代码逆向分析基础详解	刘晓阳
深度探索 Go 语言——对象模型与 runtime 的原理、特性及应用	封幼林
深入理解 Go 语言	刘丹冰
Vue+Spring Boot 前后端分离开发实战	贾志杰
Spring Boot 3.0 开发实战	李西明、陈立为
Flutter 组件精讲与实战	赵龙
Flutter 组件详解与实战	[加]王浩然（Bradley Wang）
Dart 语言实战——基于 Flutter 框架的程序开发（第 2 版）	亢少军
Dart 语言实战——基于 Angular 框架的 Web 开发	刘仕文
IntelliJ IDEA 软件开发与应用	乔国辉
Python 量化交易实战——使用 vn.py 构建交易系统	欧阳鹏程
Python 从入门到全栈开发	钱超
Python 全栈开发——基础入门	夏正东
Python 全栈开发——高阶编程	夏正东
Python 全栈开发——数据分析	夏正东
Python 编程与科学计算（微课视频版）	李志远、黄化人、姚明菊 等
Python 游戏编程项目开发实战	李志远
编程改变生活——用 Python 提升你的能力（基础篇·微课视频版）	邢世通
编程改变生活——用 Python 提升你的能力（进阶篇·微课视频版）	邢世通

续表

书　名	作　者
Python 数据分析实战——从 Excel 轻松入门 Pandas	曾贤志
Python 人工智能——原理、实践及应用	杨博雄 主编
Python 概率统计	李爽
Python 数据分析从 0 到 1	邓立文、俞心宇、牛瑶
从数据科学看懂数字化转型——数据如何改变世界	刘通
FFmpeg 入门详解——音视频原理及应用	梅会东
FFmpeg 入门详解——SDK 二次开发与直播美颜原理及应用	梅会东
FFmpeg 入门详解——流媒体直播原理及应用	梅会东
FFmpeg 入门详解——命令行与音视频特效原理及应用	梅会东
FFmpeg 入门详解——音视频流媒体播放器原理及应用	梅会东
Python Web 数据分析可视化——基于 Django 框架的开发实战	韩伟、赵盼
Python 玩转数学问题——轻松学习 NumPy、SciPy 和 Matplotlib	张骞
Pandas 通关实战	黄福星
深入浅出 Power Query M 语言	黄福星
深入浅出 DAX——Excel Power Pivot 和 Power BI 高效数据分析	黄福星
从 Excel 到 Python 数据分析：Pandas、xlwings、openpyxl、Matplotlib 的交互与应用	黄福星
云原生开发实践	高尚衡
云计算管理配置与实战	杨昌家
虚拟化 KVM 极速入门	陈涛
虚拟化 KVM 进阶实践	陈涛
边缘计算	方娟、陆帅冰
LiteOS 轻量级物联网操作系统实战（微课视频版）	魏杰
物联网——嵌入式开发实战	连志安
HarmonyOS 从入门到精通 40 例	戈帅
OpenHarmony 轻量系统从入门到精通 50 例	戈帅
动手学推荐系统——基于 PyTorch 的算法实现（微课视频版）	於方仁
人工智能算法——原理、技巧及应用	韩龙、张娜、汝洪芳
跟我一起学机器学习	王成、黄晓辉
深度强化学习理论与实践	龙强、章胜
自然语言处理——原理、方法与应用	王志立、雷鹏斌、吴宇凡
TensorFlow 计算机视觉原理与实战	欧阳鹏程、任浩然
计算机视觉——基于 OpenCV 与 TensorFlow 的深度学习方法	余海林、翟中华
深度学习——理论、方法与 PyTorch 实践	翟中华、孟翔宇
HuggingFace 自然语言处理详解——基于 BERT 中文模型的任务实战	李福林
Java+OpenCV 高效入门	姚利民
AR Foundation 增强现实开发实战（ARKit 版）	汪祥春
AR Foundation 增强现实开发实战（ARCore 版）	汪祥春
ARKit 原生开发入门精粹——RealityKit + Swift + SwiftUI	汪祥春
HoloLens 2 开发入门精要——基于 Unity 和 MRTK	汪祥春
巧学易用单片机——从零基础入门到项目实战	王良升
Altium Designer 20 PCB 设计实战（视频微课版）	白军杰
Cadence 高速 PCB 设计——基于手机高阶板的案例分析与实现	李卫国、张彬、林超文
Octave 程序设计	于红博
Octave GUI 开发实战	于红博
全栈 UI 自动化测试实战	胡胜强、单镜石、李睿